The Design
Manager's Handbook

The Design Manager's Handbook

John Eynon
RIBA MAPM FCIOB CEnv.
Consultant / Director
Open Water Consulting Ltd

CIOB

WILEY-BLACKWELL
A John Wiley & Sons, Ltd., Publication

This edition first published 2013
© 2013 The Chartered Institute of Building

Blackwell Publishing was acquired by John Wiley & Sons in February 2007. Blackwell's publishing programme has been merged with Wiley's global Scientific, Technical and Medical business to form Wiley-Blackwell.

Registered office:
John Wiley & Sons, Ltd, The Atrium, Southern Gate, Chichester, West Sussex, PO19 8SQ, UK

Editorial offices:
9600 Garsington Road, Oxford, OX4 2DQ, UK
The Atrium, Southern Gate, Chichester, West Sussex, PO19 8SQ, UK
2121 State Avenue, Ames, Iowa 50014-8300, USA

For details of our global editorial offices, for customer services and for information about how to apply for permission to reuse the copyright material in this book, please see our website at www.wiley.com/wiley-blackwell.

Library of Congress Cataloging-in-Publication Data
Eynon, John.
  The design manager's handbook / John Eynon, RIBA MAPM FCIOB CEnv., consultant/director, Open Water Consulting Ltd.
       pages cm
  Includes bibliographical references and index.
   ISBN 978-0-470-67402-4 (pbk. : alk. paper) 1. Building–Planning–Handbooks, manuals, etc. 2. Building–Superintendence–Handbooks, manuals, etc. 3. Construction industry–Planning–Handbooks, manuals, etc. I. Title.
   TH438.E98 2013
   624.068'4–dc23
                                        2012028579

A catalogue record for this book is available from the British Library.

Wiley also publishes its books in a variety of electronic formats. Some content that appears in print may not be available in electronic books.

Set in 10/13 pt Gothic by Toppan Best-set Premedia Limited

Cover design by Steve Flemming
Cover image courtesy of Shutterstock

1 2013

To my father, Kenneth James Eynon, 1925–2008
A gentle, private and resolute man who built his home in my heart

If you just want to do the 9–5, go home at night and not think about improving at DM at all, then this book probably isn't for you.

However, if you want to grow in your understanding and your role, and 'poke the box' or push the boundaries of your thinking, then you'll find some ideas here to get you started!

'Those who have aimed at acquiring manual skill without scholarship have never been able to reach a position of authority to correspond to their plans, while those who relied only upon theories and scholarship were obviously hunting the shadow, not the substance. But those who have a thorough knowledge of both, like men armed at all points, have the sooner attained their object and carried authority with them.'

– Vitruvius, *On Architecture*, 1st century BC
Perhaps the first commentator on Design Management!

Contents

Contents

Foreword

Design Management, to most, is a relatively new form of consultancy.

Its history dates back to the huge increase in Design and Build contracts in the late 1980s. During that period there was a fundamental shift away from traditional contracts, where architects ran the construction contract. D&B offered many benefits to clients, most notably a fixed price and a fixed programme. Design Managers at the time usually came from a construction background and their task was to take an existing design, already developed up to planning level, and procure the work for the best possible price with either the original architect or a new design team. However, it was often associated with erosion in terms of design quality, since finance and speed were the most important measures of success.

Over the last fifteen years there has been a significant change in the role, owing to the emergence of government-funded projects such as LIFT, BSF and other forms of PFI. These projects involved all forms of public buildings including libraries, town halls, health centres and schools amongst others. The challenge for the construction industry in these projects was that architects and the rest of the design team would now be employed from the very earliest stages by a contractor, and therefore the contractors needed to act as clients from day one.

It was in this area that the industry lacked experience. Traditional Design Managers were more used to procuring an 'oven-ready' design and lacked skills in briefing and the team building necessary to motivate consultants in a positive manner. On the other side, design teams lacked skills in understanding the link between design, cost and construction in the fast-track programming that accompanied these new forms of procurement. For the very first time, contractors, engineers and architects needed to work together from the very earliest stages if they were to win lucrative contracts. It was in this arena that Design Managers became a vital component of the team, forming a bridge between good design and clever construction. The new Design Managers came from all sides of the industry: architects, engineers and surveyors as well as contractors. They often became 'translators' for both parties, and this handbook highlights the complex set of skills needed to make the job a success.

The many well-designed public buildings procured, particularly over the last five years, are a testament to the way the construction industry has embraced these new challenges and I believe successful design management has been the vital component. Over the last few years we have seen the process become more sophisticated as the industry has learned how to improve its service to clients by embracing new technologies such as BIM, off-site manufacture and standardisation in an attempt to provide more affordable solutions in these austere years. In this sense, the role remains more important than ever.

This is the first time such a handbook has formally set out the scope and tasks necessary to perform Design Management and as such I believe it will become a 'bible' within the industry for many years to come.

Paul Monaghan is an Architect and a founding director of Allford Hall Monaghan and Morris (ahmm.co.uk). The practice was founded in 1989 and now works worldwide. Paul works on a wide range of projects, including masterplanning, arts, educational buildings, housing, offices, public buildings and health buildings. Recently completed projects under his direction include Kirk Balk Community College, Anne Mews, Villaggio in Ghana, Unity on Liverpool's historic waterfront, Latitude House, Barking Central, Westminster Academy and Kentish Town Health Centre. He is currently working on a further wide range of projects, including Nine Elms, several high-profile schools and the reinvention of Liverpool's Royal Court Theatre.

Paul has lectured throughout the UK, and has been an external examiner at several Universities, including Liverpool, Westminster and Southbank. He has been Vice Chair of the CABE Schools Design Review Panel and is currently on the CABE National Design Review Panel. Paul has been chairman of the Young Architect of the Year awards, and up until 2010 chaired the RIBA awards panel. He is also a RIBA Client Design Advisor.

Preface – 'My Why'

Welcome to the CIOB Design Manager's Handbook!

A while ago I read a book by Robert Pirsig called 'Zen and the Art of Motorcycle Maintenance'. In his book he discusses the philosophy of maintaining your motorcycle and the key ingredient that you need – 'Gumption'.

> *'I like the word "gumption" because it's so homely and so forlorn and so out of style that it looks as if it needs a friend and isn't likely to reject anyone who comes along. I like it also because it describes exactly what happens to someone who connects with Quality. He or she gets filled with gumption!*

> *'A person filled with gumption doesn't sit around dissipating and stewing about things. He's at the front of the train of his own awareness, watching to see what's up the track and meeting it when it comes. That's gumption.*

> *'If you're going to repair a motorcycle, an adequate supply of gumption is the first and most important tool. If you haven't got that, you might as well gather up all the other tools and put them away, because they won't do you any good.'*

The same could be said of the Design Manager. It is an under-sung role, not for the faint-hearted, right in the fulcrum and the heat of project delivery – and gumption is a key requirement.

An alternative title for the Handbook might have been 'Zen and the Art of Design Management', but that probably wouldn't have found favour with the powers that be!

So why the Design Manager's Handbook?

From my earliest time as a site-based Design Manager I always wanted to join up the bigger picture on DM. It is so easy to become isolated on site or in an office and not be aware of wider opinion and current thinking. For myself, I find that understanding the broader context helps me to make sense of my own role and contribution.

I think I've always wanted to connect things or make connections, because beneath the surface, life is so much more connected and holistic than we recognise sometimes, or can imagine. My aim in working with the CIOB on Design Management is to **connect** – people, knowledge and experience, to enable communication and collaboration.

> *'DM is intrinsically connected – to be effective, it has to be!'*

It also seemed to me that '*Design*' was looked upon as something of a 'black art' by many people in construction, rather like oil and water – there is this perception that they don't mix! (Or perhaps they can . . . as we shall see later!) And yet as I have talked to people from across the industry, in different sectors, different businesses and organisations, the same kinds of issue have emerged: confusion about the role, limited understanding of the scope and potential of DM, lack of recognition, profile and worth for DMers.

DM can be such an intangible proposition sometimes, but you can certainly experience the results and outcomes. There is no silver bullet here. Good Design Management is not rocket science. Simply by good planning, communication, collaboration and management, combined with effective teamwork, and the use of the right tools and techniques, it is possible to achieve outstanding results and produce real value for our customers. In addition, we can have a positive experience while we're in the process, as well as obtaining the desired end result.

This Handbook collects together much of what I've learned over the years, combined with the input of a few others who have joined with me on this journey.

The book is for all those brave souls engaged in the DM role on a daily basis: whatever type of business they're in, site-based or in an office, standing in that 'fuzzy' zone between design and construction. It is written more from a contractor's perspective, as that represents most of my experience now. However the processes, tools and concepts should be familiar to all, whatever seat round the project table they currently occupy.

It is my aspiration that this Handbook, together with the other thrusts of the CIOB Design Management project, will draw some threads together, *'connect'*, and enable Design Management in construction to come of age.

As we continue on this journey, I want to give everyone involved in DM some context and focus, as well as providing a platform upon which they can work, be understood, grow their roles and careers – and as a consequence to firmly plant DM on the construction sector map.

It is my hope that the CIOB DM project will give focus to everyone involved in Design Management across the industry, and enable the discussions to take place and the questions to be heard.

My experience is that most people involved in DM love to share ideas, communicate, collaborate and innovate. They are passionate about what they do.

My aim is to get people talking together about DM, sharing their experience and ideas on how to do things better than the last time around. The results of such communication and collaboration will be truly exciting, and will make a significant contribution to taking our industry forward. And to be honest, I've had a great time doing this, discussing DM with people across the industry, exchanging ideas and discussing how can we do things better and take the industry forward.

In writing the Handbook, and working with many people discussing Design Management over the last few months, I have been awestruck by the rich diversity of opinions and ideas – there is no vacuum, and there is never a short discussion! This Handbook is the first step in enabling that wider discussion and exchange of ideas to take place, and providing a focus for debate. In time I hope it will develop into a consensus on the activity and the role, as well as providing a best-practice model for DM within our industry.

John Eynon
Brighton
March 2012

Acknowledgements

As with any project, this has been a team effort. Many people have had an input or influence on the development of the Handbook in many different ways, and here I can mention only a few.

The Board of the Faculty of Architecture and Surveying of the CIOB, particularly John Hughes, Simon Matthews, Craig Hatch and Bob Giddings. They believed in the idea of a DM project at the start and kept the momentum going forward.

Asmau Nasir-Lawal of FAS, who has worked with me on the group and also the DM conference.

Saleem Akram at the CIOB, who put me on the spot, pushed me in the right direction in producing the Handbook, and has provided ongoing support.

My 'A team' at the CIOB: Sarah Peace, Veronica Dunn and Una Mair, who have provided exemplary admin support, guidance and encouragement and kept me on the straight and narrow. Sarah has been a good friend, mentor and guide throughout the entire process of bringing this book together – without Sarah, we simply would not have got here!

I'd also particularly like to mention my old 'governor' at Carillion, Paul Shadbolt; we discussed design management all the time, and looking back I can see that Paul triggered a lot of ideas that have come together in the CIOB project.

John Rivett, who over my time in DM has been both a good friend and a mentor.

Alec Newing, who has helped 'tend the DM garden' with me over the last few years, and kicked around all sorts of ideas.

Alan Mossman, who has made me pause and think on many an occasion, but always for the good, and who has made several valuable contributions to the Handbook.

Saima Butt, for her perceptive insights into how as human beings we contribute to the project process and also for our many thought-provoking discussions.

Michael Graham, who has provided some distinctive perspectives on the value of design and the value-management process.

My wife Anne-Marie, who has lived with the ups and downs of putting this book together for some considerable time, but has given me the space and support to do it.

Although I don't know them personally, Seth Godin and Simon Sinek; I have found their blogs and writings truly inspirational.

And finally for this section – to colleagues and friends at Carillion and Wates and to all the DMers, designers and constructors, and everyone else I have worked with in the last 30 years. In some way you have all shaped my thinking, which has somehow found its way into this Handbook.

Special thanks to guest contributors

The Handbook and appendices include a number of contributions by others, which I have included because they are topics of interest about how our industry is developing and because they might form jumping-off points to provoke further research and learning by DMers reading the Handbook. Special thanks to them all for their time, and the effort and ideas they have contributed.

Contributors
Paul Monaghan – the Foreword
Alan Mossman – Last Planner in Section 2, and Appendices A and C
Alec Newing – input on cost management, Section 2
Michael Graham – input to Value Management Section 6, and Appendix D
Paul Waskett and Andrew Newton – Appendix B
Paula Bleanch – input to Training, and Appendix F
Saima Butt – input to People and Appendix E.
And also Gerard Daws of NBS Schumann Smith for his case studies.

CIOB Working Group

As part of the CIOB project on Design Management, we have had a working group 'chewing the fat' over various aspects of DM. Many thoughts and ideas from our several meetings have found their way into the Handbook, and for that I am thankful for everyone's input and generosity with their time. The discussions were challenging, stretching, thought-provoking, but always positive and constructive, and they brought home to me the diversity of opinion, and the depth and breadth of the impact that DM has had on the industry, businesses, projects and people.

CIOB DM Working Group members

John Eynon	FAS Board (Chair)
Asmau Nasir-Lawal	FAS Board
Alec Newing	DM Professional
Anthony Kelly	University of Greenwich
Chris Allen	Vinci Construction Group Plc.
Diane Dale	CIAT
John Rivett	Wates Group
Keith Snook	CIAT
Michael Graham	UK Value Management
Paula Bleanch	Northumbria University
Philip Yorke	DM Professional
Sam Allwinkle	CIAT
Sarah Peace	CIOB
Scott Fenton	3D Reid
Stephen Emmitt	Loughborough University
Tahar Kouidar	Robert Gordon University, Aberdeen
Tara Pickles	CIAT
Una Mair	CIOB
Veronica Dunn	CIOB

Note to the reader

The Chartered Institute of Building, CIOB, has various Faculties, one of which is the Faculty of Architecture and Surveying (FAS). Over the last few years on FAS we have been discussing Design Management, and out of these discussions has arisen the Design Management project:

The CIOB DM project has four main objectives:

- to produce The Design Manager's Handbook

- to hold a DM conference

- to develop a learning and development framework for DM

- to facilitate the formation of a cross-industry SIG for Design Management.

For further information, contact the CIOB.

Abbreviations

From the Preface and Acknowledgements you can see that I am using a few terms and abbreviations you may not be familiar with.

For ease of reference I am using two simple abbreviations shown below:

- DM – Design Management

- DMer – Design Manager

People involved in DM will all know that the roles and job titles can vary considerably. As you are reading this you could be a Design Manager, Design Co-ordinator, Technical Manager, BIM Manager or Co-ordinator, Design Leader, Project Manager, Pre-construction Manager, Bid Manager, Construction Manager, Surveyor and so on. You might be in an office or working on site, for a contractor, sub-contractor, consultant or as part of the customer's organization. So for the purposes of this discussion I am using Design Manager, DMer, as a generic title – the person who takes the lead on Design Management activities.

There is a glossary of other terms provided below:

AEC	Architecture Engineering and Construction
AI	Architect's Instruction
AIA	American Institute of Architects
AM	Asset Management
APM	Association for Project Management
B555	BSI Committee for construction, design, modelling and data exchange
BIM	Building Information Modelling
BIS	Government Department for Business Innovation and Skills
BRE	Building Research Establishment
BREEAM	BRE Environmental Assessment Method
BSI	British Standards Institute
BuildingSMART	International Alliance for Interoperability
CAD	Computer Aided Design
CADCAM	Computer Aided Manufacture
CAFM	Computer Aided Facilities Management
CAWS	Common Arrangement of Work Sections
CC	Construction Confederation
CDE	Common Data Environment
CDM	Construction Design and Management regulations
CDP	Contractor's Design Portion (sub-contract design package/works package)
CfSH	Code for Sustainable Homes
CIAT	Chartered Institute of Architectural Technologists
CIBSE	Chartered Institute of Building Services Engineers
CIC	Construction Industry Council
CIMM	TfL programme and project process for LU
CIOB	Chartered Institute of Building

CI/SfB	UK version of the Construction Indexing Classification system for construction products
CITE	Construction Industry Trading Electronically
CO2	Carbon Dioxide
CO	Change Order
CO2E	Carbon Dioxide Equivalent
COBIE	Construction Operations Building Information Exchange
CPA	Construction Products Association
CPD	Continuing Professional Development
CPIc	Construction Project Information Committee
DEC	Display Energy Certificate
DECC	Government Department of Energy and Climate Change
DFMA	Design For Manufacture and Assembly
DM	Design Management
DMer	Design Manager
DMS	Document Management System
EDMS	Electronic Document Management System
EPC	Energy Performance Certificate
ERG	Efficiency Reform Group
ERM	Exchange Requirements Model
FM	Facilities Management
GCCB	Government Clients Construction Board
GCCG	Government Clients Construction Group (UK)
GHG	Greenhouse Gases
GIS	Geographical Information System
GRIP	Network Rail project management process
GSA	General Services Administration (USA)
IAI	International Alliance for Interoperability (BuildingSMART)
iBIM	Fully integrated BIM
ICE	Institution of Civil Engineers
ICT	Information and Communications Technology
IDC	Integrated Design and Construction
IDM	Information Delivery Manual (BuildingSMART)
IFC	Industry Foundation Classes (BuildingSMART)
IFD	International Framework for Dictionaries (BuildingSMART)
IGT	Innovation and Growth Team – UK Government
IP	Intellectual Property
IPD	Integrated Project Delivery
IPI	Integrated Project Insurance
IRS	Information Required Schedule
ISE	Institution of Structural Engineers
IUK	Infrastructure UK
JCT	Joint Contracts Tribunal
LEED	Leadership in Energy and Environmental Design assessment methodology
MVD	Model View Definitions
NEC	National Engineering Contract
NIBS	National Institute of Building Sciences (USA)
NRM	New Rules of Measurement (RICS)
OGC	Office of Government and Commerce
Omniclass	Classification system for building components (USA)
RIBA	Royal Institute of British Architects
RICS	Royal Institute of Chartered Surveyors
RFI	Request For Information

SME	Single Model Environment
SMM7	Standard Method of Measurement Edition 7 (RICS)
SMP	Standard Method and Procedure
Spearmint	TfL programme and project management process for surface transport
SPM	Single Project Model
Tfl	Transport for London
TQ	Technical Query
Uniclass	Classification system for building components (UK) base on CAWS , CI/SfB and other systems

1

Introduction

The CIOB Design Manager's Handbook

Design – 'the art or action of conceiving of and producing a plan or drawing of something before it is made'.

Management – 'the process of dealing with or controlling things or people'.
— Oxford Dictionary of English

Introduction

For as long as humankind has existed we have been making things. Someone has an idea, and then someone makes it. Different people might be involved at different stages.

This process of conceiving an idea, designing, planning and making something a reality has always been managed by someone, perhaps at first only intuitively. The master-builders of old had to provide information in some form at the right time for the craftsmen to carry out their work. The Pyramids, the Pantheon, the Acropolis, St Paul's Cathedral – all required a designer, but also needed someone to convey and translate that information over to the craftsmen and builders.

Where there is a design process happening, a management process needs to be happening in parallel to enable the design to reach fruition successfully. However, sometimes this does not happen, because there is a vacuum through lack of recognition of the issues or through a lack of understanding or expertise.

There is a whole area of discussion around the necessity for a discipline or defined role of the 'Design Manager'. Does a DMer add value to a project? Is this a new discipline, or is it a part or a subset of an existing skill set?

When design is happening as a process, then **Design Management as an activity needs to take place**. I will leave the discussion about who should do it until a little later!

I have found that there is too much focus on roles. If instead in looking at these we focus on activities – **'What needs to happen?'**, **'Who is then best placed to do it?'** – that gives clarity to structuring an approach.

This Handbook is the culmination of several years' work within the CIOB, principally within the Faculty of Architecture and Surveying. It is specifically written for those brave souls actually carrying out the DM role within the Construction and Built Environment Sector. You might be based on a construction site, or in an office working on tenders or pre-construction stages, or in a designer's office or perhaps as part of a customer's in-house team. You might be a young graduate just starting

The Design Manager's Handbook, First Edition. John Eynon.
© 2013 The Chartered Institute of Building. Published 2013 by Blackwell Publishing Ltd.

out on your first job, or someone much older and more experienced. Most of the terminology and perspective I have used derive probably more from a contracting viewpoint, which reflects my experience, so I make no apologies for that. Whatever your role is in the project team or that of your organisation, I am sure you will bump into a lot of this on a daily basis!

Nevertheless, whatever stage you're at in your career, this Handbook is for you. I hope you will find it of use as an aide-memoire and that you will dip into it to see what the growth and development areas for DM in the future might be, as well as finding some ideas that will help in your daily job.

So . . . what is Design Management?

Definition

The CIOB DM working group, after much discussion, settled on the following working definition:

A definition of Design Management – The Activity

Design Management includes the management of all project-related design activities, people, processes and resources:

- Enabling the effective flow and production of design information

- Contributing to achieving the successful delivery of the completed project, on time, on budget and in fulfilment of the customer's requirements on quality and function in a sustainable manner

- Delivering value through integration, planning, co-ordination, reduction of risk and innovation

- Achieved through collaborative and integrated working and value-management processes.

We need to recognise that all sorts of people now populate the territory of 'Design' – traditionally the realm of professional designers such as architects, technologists, and structural, civil and MEP services engineers. However, we have seen the rise of specialist subcontractor designers, particularly in the fields of cladding, building envelope design, MEP building services and latterly environmental technologies. Also there are the specialist consultants, such as those for planning, acoustics, environmental matters and so on. The finished design of any building project will be the result of the efforts of possibly hundreds of people involved at various stages. All of these people need leadership and co-ordination to achieve the end result. This is the stage on which DM performs.

So who is the Design Manager?

Traditionally the architects would have seen themselves as natural leaders of the project team, organising, co-ordinating and administering the contract – taking the brief from the customer, organising the team and leading the design and contract processes through to completion. However, as we know, on all but the simplest and smallest of projects, the process is no longer that simple!

Procurement of building projects has become more complex and technically demanding. It seems that at the drop of a hat, another specialism appears. The Quantity Surveyor, Project Manager, Construction Manager, CDM-C, Planning Consultant, Party Wall Surveyor and others all have to be integrated into the project delivery process. In this context, is DM simply just another 'discipline'?

The role and activity of Design Management has gained ground and risen in profile, particularly through the widespread adoption of Design and Build-style procurement, and the rise of the specialist subcontractor. Is it significant that over this time span, as the education of the architect has moved into the realm of academia, the need for the DMer has emerged? Could there be a link? As a consequence of this shift in education and training, the architect's influence on the design and construction processes has declined. The role of architects has become distanced from the actual process of building through the education process and a lack of practical experience and knowledge. Looked at in this way, the disconnect between design and construction is revealed and, strangely, it is precisely this 'gap' that the Design Manager can fill.

The professional designers such as architects and engineers are now incredibly reliant on specialist subcontractors to realise their visions. Much of the procurement process is about selection of the right supply-chain members with the necessary expertise to complete the design for manufacture, installation and construction. The supply-chain resources and expertise are marshalled by the main contractor, and within their team will probably be a Design Manager helping to manage and co-ordinate all these inputs.

Consumer contractor

Several years ago, on a large commercial building project in central London, the client's project manager explained it to me thus:

> *'As the main contractor, you are the consumer of the design information produced by the design team. Therefore it is appropriate that you (i.e. me!) manage the process.'*

So from that time on, I chaired all the meetings, produced the minutes, and followed up on the actions – he had a point!

Irrespective of how a project starts, the execution and delivery of the building phase will be led by the contractor, who carries most of the attendant risk. There is a strong argument for the contractor dictating the nature of the design process to enable information to be produced correctly and on time, and to ensure that the procurement and construction processes proceed as efficiently and economically as possible. I know that some designers will disagree, but rarely are designers exposed to anywhere near the magnitude of commercial risk to their business that contractors generally carry in the delivery process.

Perhaps the jury is still out on the matter, but I think over the last decade the pendulum has swung decisively in this respect. Most major projects are now delivered by contractor-led teams, supported by an army of specialists including designers and subcontractors. We have seen this particularly on public-sector frameworks and with major commercial-sector clients. I suspect it will be a long time before the pendulum swings back, if it ever does. Clients require certainty of delivery, in terms of time, cost and quality. Early engagement of the contractor in the process is one of the main ingredients in providing this certainty and reducing the client's risk.

Project circumstances vary, but probably for a large proportion the DM role sits with the contractor. Whether this is a defined role, or part of, say, a project manager's brief, is another discussion; but the role and the activity need to take place for the project to be successful. Certainly anecdotally, probably most 'design managers' in the UK are within contracting and supply-chain organisations.

Production of design information that is accurate, co-ordinated, buildable and in line with the budget is a mission-critical activity – so it is an activity that is probably best left in the hands of someone who understands the processes, the requirements and tools needed to deliver successfully. If the design fails at any stage for whatever reason, be it poor quality, late, over budget, etc., then the results of that failure will impact on the project delivery, sometimes with catastrophic effect.

Design quality

Detractors of contractor-led delivery will cite the dumbing-down of design quality for design-and-build contracts. That opinion might have held some weight several years ago, but now there are numerous examples of high-quality buildings delivered through Design and Build-style procurement with leading architects as part of the team, and also those projects are winning design and industry awards in the process.

Sadly there are still exceptions, and usually they are the result of conflicting project drivers for the scheme. What is most important to the client? Finishing early, being under budget, or having a building designed with a world-class signature architect? In defining the parameters for delivering the project, sometimes something has to give and frequently the contractor, after a hard-won tender process, is the bearer of bad news on the cost plan!

Most major contractors are now exemplary in their approaches to corporate social responsibility, sustainability, the environment and health and safety. Indeed, they have to be, as these are frequently part of the scoring criteria for awarding contracts.

Where contractors perhaps need to step up and catch up with their design brethren is in articulating and understanding their impact on the built environment, with regard to the quality of the urban fabric they leave behind. If contractors are to credibly lead the team, transforming our towns and cities and the environment, then they need to understand urban design and consider the bigger picture.

If up to 70 per cent of the buildings that we are constructing now, together with those already existing, will still be standing in 2050 and beyond, what sort of legacy are we leaving to future generations? What impact will the homes, workplaces, civic buildings that we produce now have on people in the future and their daily lives? Are we making life better? Do we care? We all play a part in this realm of

> *Is it enough to say: 'We did what we were asked/instructed/told to do,' and walk away at completion?*

the built environment, no matter what role we have in the project, and therefore we have to take responsibility.

It is these sorts of issue that intelligent contractors should be able to discuss openly with architects, planners and urban designers. Perhaps an indication here would be to say we have achieved 'working towards' on this aspiration! The DMer with a foot in both camps is able to help in taking this aspect forward.

The DM role

As a construction role, DM is a relative newcomer, probably gaining prominence in the 1990s and arising from the need for better co-ordination and delivery of design information from design teams to main contractors – particularly on design-and-build contracts and also on CDP-style procurement routes (*CDP – Contractor's Design Portion – is a subcontract works package with specialist design responsibility*). The increasing use of contractor-designed packages using specialist subcontractors has also increased the demand for the role, particularly as related to MEP services, structures, cladding/envelope solutions and other specialist areas.

Working with BIM increases the need for co-ordination, and management of design process and collaboration. Some will argue that BIM will solve a lot of our problems in this area. However, it is only a tool, operated by people – people who still need to produce the right information and inputs and complete their activities at the right time for BIM to be effective, which is DM in action.

Any building project of a reasonable size is the result of the work of hundreds of people both on and off site, ranging from designers, engineers, subcontractors, specialist consultants and suppliers and manufacturers, and – owing to global sourcing – probably from around the world. Never has the need been greater for a facilitator of the integrated and co-ordinated design process.

The DM role encompasses managing internal and external design consultants, controlling the development of the design concept into manufacturing and installation and construction status information. This type of role also exists in other industries besides construction, such as branding, manufacturing and industrial design, software development, automotive, petrochemicals and space engineering – essentially any process where design is involved in producing a finished, built or manufactured product. It is about process and people management, requiring a blend of technical and commercial awareness, mixed with project management and people and communication skills.

All leading UK contractors have design management teams. Management consultants sell design management services, and some design consultants, have their own in-house design managers to help with their design processes – effectively project managers of design.

DM in its broadest sense also includes management of mechanical and electrical plumbing (MEP) services design, which is mainly carried out through specialist MEP contractors. MEP packages represent a significant proportion of the content of most building contracts – typically as much as 25–35 per cent

Owing to its position in the design and construction processes, DM sits very closely with other roles in project teams, such as bid management, estimating, pre-construction management, and planning/programming. This includes being 'customer-facing' – for instance, dealing with briefing, managing compliance with the employer's requirements, and communication relating to design issues and progress.

Site-based DM roles are usually much more focused on construction information, maintaining and co-ordinating the flow of design information to enable construction work to proceed efficiently, as well as facilitating cost control and constructability review processes.

So that is a whistle-stop tour of the territory: but what about the person carrying out the role?

The Design Manager

People enter this field from all sorts of backgrounds, including architects, technologists, engineers, surveyors, administrators, construction managers and site managers. Qualifications in the role vary from none, to construction and design degrees or BTEC/ONC/HNC. There are now a few DM degree courses available, for example at the Universities of Greenwich, Loughborough, Northumbria and Robert Gordon.

Just as it is not easy to categorise people in the DM role, so job titles are equally diverse, although the actual role can be essentially the same (but not always!). Examples include Design Manager, Design Co-ordinator, Technical Manager/Co-ordinator, Building Services Manager, MEP Manager, Pre-construction Manager (which might be the same role or a different one!); they could even include Bid Manager or Project Manager, with a DM brief as part of their role. Career grades could be Assistant, Senior, Principal, Lead, and possibly in some cases Director. The increasing popularity of BIM has now brought us the BIM Manager, Co-ordinator, Leader, etc. – perhaps the Design Manager of the future?

To add another layer of complexity, every business has a slightly different approach as to how their project and business structures accommodate DM, as well as variations in the processes and procedures they follow. But what about the people themselves? What skills, knowledge and experience do they need?

The role

Their competency set could cover (in no particular order of importance):

Technology

- Basic understanding of the industry and how it works
- Building technology and systems, including structures and MEP
- Codes of Practice and Regulations, e.g. Planning, Listed Buildings, Building Regulations, BS/EN standards
- Design quality and urban impact
- Health and safety, including CDM
- Sustainability and the environment.

Process

- Construction process and logistics
- Design process
- Management tools and strategies
- Planning, programming, time-management techniques.

Commercial

- Commercial/cost/estimating
- Contract, legal
- Procurement routes.

Personal

- Dealing with people, communicating, influencing, negotiating, managing
- Presentation, communication
- Telepathy (so you can actually understand what people mean, rather than what they say) and necromancy (so you can predict the future, and avoid all those mistakes that you should have known about beforehand) – *I'm just checking that you're paying attention!*

And the list could go on!

After examining several job profiles for Design Management roles, I feel that the DMer needs to be some kind of Superwoman or Superman – there is an element of being all things to all people and knowing everything about everything! In a typical week, you could be in front of the client, meeting the local planning officers, leading a design team meeting, meeting subcontractors to discuss details and then walking round the site to check on built work for compliance with the design information. So it can be extremely varied, and the mindsets, knowledge and skills required in these different situations are as equally diverse. It's quite a challenge on a personal level for the Design Manager, let alone the technical and business knowledge involved.

At first sight all this could seem quite daunting and challenging, and yet there are so many positive aspects to the DMer role: being part of a team; developing relationships that sometimes can last a lifetime; becoming knowledgeable about all aspects of the design and construction processes; always learning; seeing a project come together in design, and then going out to watch it being built; and the thrill of handing over the finished building to a satisfied customer. It's definitely a role that is challenging, but also extremely fulfilling, with never a dull moment and I think a great future – something to aspire to!

The person

Let's develop this further; let's think about how people 'occur'. Our personalities, moods, personal agendas all affect how we work, collaborate, communicate or not, as the case may be. I am sure that we have all experienced personality clashes and difficulties in our lives; then sometimes, in other situations, it is no effort at all to work with certain people. Awareness of self and of others is another strand to this equation.

Depending on your entry route into the construction industry and your qualifications and background, some of the above competencies listed will be covered, but there will still be gaps elsewhere. If you have come through the trades, then you might not have a great deal of experience of working with designers. Similarly, as a designer you might not have spent much time on site, understanding how buildings really come together physically and how site-delivery teams work.

It's no surprise, therefore, that there is some confusion around the DM role. Different businesses have different emphases, the role and requirements are not universally understood, and the skills, qualifications and experience of the people actually doing Design Management represent a hugely diverse spectrum.

Hallmarks of DM

For many years I have held the belief that there are generic hallmarks and principles of good Design Management. These hallmarks set the framework for how DM should work, and as a result they scope the role of the Design Manager. Also, the different challenges that the many stages of the design and construction processes pose to the DM are significantly different, requiring appropriate and relevant skills and understanding and differing mindsets.

Is it realistic to expect someone to excel at every stage of the design, procurement and construction processes? As individuals we naturally gravitate to those stages of the process that we are best at, are interested in or even passionate about. And this is not just about construction-oriented people: in design practices also, there is recognition of interest and ability, allowing people to specialise at different stages of design, ranging from concept to detail and delivery, and in some cases business administration and contract.

Stages of DM

Pre-construction and site delivery can be different worlds, requiring different skill sets and mentalities. Pre-construction is a world of possibilities, options and alternatives, while construction is very much about certainty of delivery – following the delivery process logically, step by step, to complete the project. The very earliest stages of a project can be just as different again – it's the world of feasibilities, business cases and value-management options. For someone who is more com-

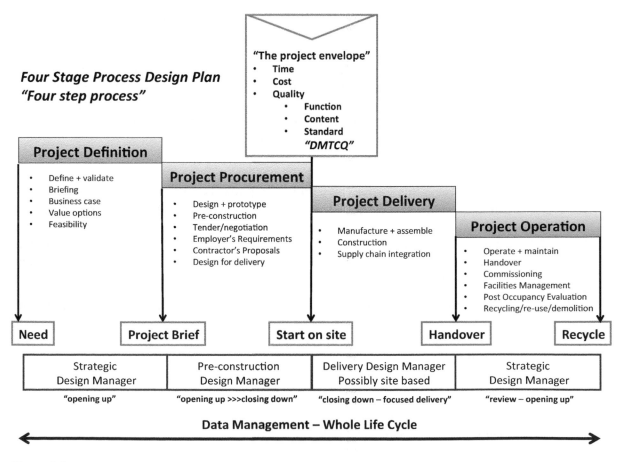

Figure 1.1

fortable with the building stage, just wanting to get on and build something, this can be a scary place! Within these broader fields Design Managers can also become expert, because of their training or experience in cladding, structures, or renewable technologies, for example.

Therefore I think it is realistic to recognise different flavours of DM or **modes** of operation; refer to the Four Stage Process in Figure 1.1. The DMer role changes in nature from stage to stage.

Design Management operates in:

1. Project definition

2. Project procurement

3. Project delivery

4. Project operation

The commentary on these four stages is as follows, and sets the scene for the discussion of the role of the Design Manager in Section 2: Process.

Project definition

What does the client want? What do they need? This could involve establishing the business case, pinning down the brief, perhaps managing the production of options or feasibilities and their review and assessment; then defining what the project is actually going to be, and the parameters for its delivery, such as the intended timescales, budget limits, and the facilities and standards to be provided.

Project procurement

Once the project direction has been established, then the design is developed for submission for planning approval and further developed for Building Regulations approval. There may well be the need for further value-management exercises to be held to ensure the project is within budget and delivering what the customer wants – but that will usually be within the now-established project parameters. Sometimes it will become necessary to go back in the process, because requirements change, budgets are exceeded or other external factors apply. Other activities here could include early supply-chain input, review of key aspects and risk mitigation. The objective of this stage is to put in place all that is required to enable delivery of the project in the next stage.

Project delivery

Once the scheme is defined, and approvals obtained, the project moves into delivery mode, with the contractor having been appointed through a tendering process, which will have established the boundaries for the project in terms of time, cost and quality (function, content and standard). The objective of this stage is to feed the construction machine the information it needs to complete the project according to programme and budget. This involves the integration and co-ordination of all the stakeholders, including designers and the supply chain. Programming, cost, 'buildability' and compliance issues all come into play, and at this stage the site-based Design Manager has to deal with them all.

Project operation

During the completion stage systems are commissioned, the project is handed over, defects are rectified and the project moves into operational mode. Training may need to be given to building managers and the FM team, and potential end users. Aftercare is becoming increasingly important in the project delivery. There are opportunities to enhance the customer's experience at this stage, as well as to develop lessons learned for future projects through Post-Occupancy Evaluation. In addition there will be ongoing operational evaluation of the project in use – say, of the adequacy of the facilities, energy and resource use and carbon emissions, for example.

Eventually the project may require alteration, extension, or perhaps demolition and recycling. Strategic input from a suitable Design Manager will assist in this process and so the cycle begins again (or ends for this particular project).

Activities, not roles

Note that I have described above the activities that need to take place. Time and expertise allowing, these activities could be carried out by a Project Manager, Bid Manager or Pre-construction Manager, as part of their skill set, for example. In terms of site delivery, the DM function could be fulfilled by a Package or Site Manager responsible for the construction of the design elements involved. Cladding, MEP service, structures would be good examples of where a suitably able and skilled Package Manager would deal with the package from design through to final completion as part of their responsibility. This could equally be applied to any aspect of the construction, once the overall parameters have been set.

The CIOB Design Manager's Handbook

So, having sketched out the field we are going to look at, this is how the Handbook is structured and the areas it will cover. It is worth keeping in the back of your mind the Four Stage Process (Figure 1.1) and DMTCQ (Section 3). These set the principles and the overall context of the Handbook.

The Handbook sections cover:

- **Process**
  Consideration of the project context and timeline, looking at some key strands of the overall process, and examples of leading industry processes, and other aspects.

- **DMTCQ**
  Guidance on the key DM aspects that form good practice for Design Management. A framework of principles that the DMer can use to manage and also analyse DM activities.

- **Tools**
  Identification of some generic tools that can be used to control design development and information production. Some simple examples and ideas on key tools to manage design successfully.

- **Procurement**
  How DM process requirements flex with procurement routes, discussion and observations. Different contract forms generate different drivers, which affect the DMer in practice. Discussion around the Employer's Requirements, Contractor's Proposals and Novation.

- **Value and innovation**
  The relevance of value management to Design Management. How VM can benefit the project in practice through Design Management. Some thoughts about how the DMer can be a catalyst for innovation.

- **People**
  Perhaps a neglected aspect is the person – how we interact, factors affecting our behaviour and the roles we take up in situations. Exploring how aware we are of the people element in the way we work.

- **Training**
  How do people find their way into DM roles? An overview of academic, professional, trade and off-the-tools routes.

- **Quality**
  Consideration of project quality, leading to a discussion around design and urban quality. Is this the province of designers only, or do contractors need to step up?

- **Stories**
  Case studies of where DM has added value and also stories of people in DM actually doing it. Their background, education, how they started out.

- **Future**
  Potential changes and trends in the industry, and impact of external factors such as climate change, resources use and demographics. Impact of BIM and technology on the role and the process.

- **Some contributions from CIOB DM working-group members interested in DM or related subjects**
  Contributions on process, value management, lean planning, tools, people skills, education and further resources for learning.

Conclusion

Just as in life, there is a connectedness about DM that means that strands intertwine and overlap. Successful DM takes in all sorts of factors: time, cost, quality, sustainability, environment, safety, logistics, buildability and so on. If there is such a thing as the 'Zen of Design Management', then it comes from the holistic harmony of all these factors resulting in the whole process flowing efficiently and productively, with no waste – perhaps rather like the concept of flow in lean construction.

However, a successful project itself is the result of many inputs and activities, and DM is but one subset of these. So for the purposes of this Handbook, focusing on inputs and areas individually for DM is necessary, but inevitably there will be some overlap and repetition due to this interconnectedness of strands and issues.

In conclusion it is difficult to say what the future might be for DM, but there is no doubt that it is here to stay, and already there are clues as to the next evolutionary step.

Remember: Wherever 'Design' is happening, DM will need to take place for the project to be successful. The questions will always be: who should do it and who is best placed to do it?

The increasing complexity of design and delivery models cries out for a leader/co-ordinator/integrator. BIM and technology will also have a huge impact. Who knows where this may lead? Today's Design Manager could be the next generation's BIM manager or technologist, dealing with compiling nD BIM that provides all design, manufacturing and construction information, as well as financial and procurement information, while still providing an ongoing vehicle for controlling carbon emissions, energy consumption and facilities management, and in addition acting as a lynchpin for design facilitation and collaboration.

'May you live in interesting times . . .' as the Chinese proverb goes, and we certainly do! There is no doubt in my mind that the rate of change and innovation in our lives is increasing. In my own lifetime so far I have seen mobile phones shrink from the size of a concrete block to almost a credit card, whilst personal computers have gone from covering the desktop, to being capable of being carried around in your pocket. It is a cross between James Bond and 'Thunderbirds', and it is all on my iPhone! Who was it at IBM that said no one would be interested in personal computers?!

For the design-and-construction industry, the next ten years will be even more exciting and challenging than anything we have seen before. Technology; the impact of Generation Y with their ingrained networking and IT skills, as they move forward into leading our businesses; the silver surfers living and working longer; the subcontractors becoming even more expert and specialist; plus the economic environment, as well as the immediate challenge of climate change. All these factors and more are going to drive the transformation of the way we work and the buildings that we design and construct.

Construction is an amazing industry in which to work. The opportunities are diverse and immense. Every day great things are accomplished in the UK and all over the world by people just doing their job, in design offices, on site, and in supply-chain workshops.

We all know the immensity of the challenges that the industry faces; they confront us every day. The solutions will not be delivered by one discipline, one institute or one sector of the industry on its own. It is only by working together, collectively and collaboratively, that these challenges and more will be met and overcome. We are all part of the problem. Conversely, we are all part of the solution.

This is totally in the spirit of Design Management.

BIM will have a huge impact on our industry. It will eventually prove to be a generational step change. It will impact on all aspects of the design, construction and delivery processes and will fundamentally affect the role of the Design Manager, bringing DM to the very centre of project collaboration and integration.

My experience of people involved in design and construction is that that they willingly share and talk about what they do, and are open to other people's ideas. Certainly my experience of Design Managers very much supports that notion.

> The DM role and activity is about bringing ideas together, connecting, integrating, communicating, innovating and collaborating.

And finally for this section

As can be seen from all the previous discussions, the world of Design Management in the construction industry is wide, diverse, far-ranging, complex and multi-layered.

On the board of the CIOB Faculty of Architecture and Surveying we have been discussing this for some time. Out of that discussion has arisen the project to develop the CIOB approach to Design Management and to provide some focus. The production of this Handbook is the first step in the process and I am hoping it will be relevant and useful and will contribute to developing DM in this great industry of ours in the future.

Whatever your role, site-based or in an office, working for the customer, designer, contractor or subcontractor, the Handbook is for you. It is based mainly on UK principles, but from an international perspective; aside from differences related to local code and compliance issues, process remains largely the same. Differing procurement and contract models mainly hinge upon the degree and the stage at which the main contractor becomes involved in the process.

This Handbook provides a focus on who and where we are, what we do, where we could go and how we might get there. My hope is that this will contribute to raising the profile of Design Managers, and getting DMers talking, collaborating, innovating and exchanging ideas across our industry. Just as importantly, it is to assist you in whatever role you face on a daily basis.

Connect – integrate – communicate – innovate – collaborate!

Let's go!

Sections 2 and 4
Brief Introduction and Contents

'Projects don't go wrong; they start wrong!'

– Anon.

'Insanity: doing the same thing over and over again and expecting different results.'
– Albert Einstein

Cheshire Puss asked Alice: 'Would you tell me, please, which way I ought to go from here?' 'That depends a good deal on where you want to go,' said the Cat. 'I don't much care where,' said Alice. 'Then it doesn't matter which way you go,' said the Cat.
– Charles 'Lewis Carroll' Dodgson
Alice's Adventures in Wonderland, 1865

'I love it when a plan comes together!'

– Col. John 'Hannibal' Smith
The A Team

Sections 2 and 4, Process and Tools respectively, form the heart of Design Management and also therefore the heart of this Handbook, making up the largest sections.

> Section 2 – Process is about: getting the answers to the Why, What, Who, When?
>
> Section 4 – Tools is about: supplying the How?
>
> This is the 'engine room' of DM.

This is where connections are made, communication happens, information is circulated, workshops are held and information is reviewed, to ensure the project is delivered on time, on budget and to the right quality – with innovation, value management and sustainability thrown in for good measure. As a DMer in your role, you will draw on resources, processes and tools to suit the circumstances and the outcomes that you need to achieve at each stage of your project.

No one industry process or tool in itself will necessarily fit all of your needs at any given time, so you will adapt and use them as required to suit your objectives and the requirements of your business and the project team.

In these two sections I have highlighted a number of processes and tools that are generally accepted across the industry and are in common use. Most contractors

The Design Manager's Handbook, First Edition. John Eynon.

will use the tools I have described in some shape or form, even if the titles and application vary slightly. For ease of reference I have included below a mini-list of contents, so that you can see what is covered and contained in each section.

2 – Process

- Introduction
- Get some 'TCQ' – Time, Cost, Quality in DM
- Defining the landscape
- DM Definition – the activity
- Supplementary definitions
- Definition commentary
- Four stage process
- Roles
- Titles
- Starting and finishing
- The overall project context
- Where is the value?
- Leveraging value
- Integrating the team
- Process models
- BIM, process and the future
- BS7000: Part 4
- BS1192: 2007
- CIC Scope of Services 2007
- The RIBA Plan of Work
- Salford Process Protocol
- Office of Government Commerce – The OGC Gateway™ Process
- Avanti
- DM and CDM – Health and Safety
- DM and Cost Management
- Last Planner
- Summary.

4 – Tools

- Introduction
- Animations, fly-throughs
- Appointment documents
- Audits
- BREEAM

- BSRIA Framework for Design Services
- Building Regulations
- Carbon emissions and energy modelling
- Code for Sustainable Homes
- Contractor's Proposals (see Section 5)
- Deliverables Schedule
- DM Project Plan
- DQI – Design Quality Indicator (see Section 9)
- Employer's Requirements (see Section 5)
- Fire engineering
- Information Required Schedule
- Interface matrix
- Matrix of Package Responsibilities
- Meetings
- Models, virtual, physical
- Peer reviews
- Planning (Development Control)
- Programmes
- Reports
- Resources schedules
- Risk assessments/schedules
- Samples/Benchmarks/Prototypes/Mock-ups
- Schedules of services
- Scope documents
- Simulations
- Stage reviews
- Status of information
- Teams
- Tests
- Tolerance/Movement schedule
- You yourself
- Summary.

2 Process

The CIOB Design Manager's Handbook

> A man is driving in his car trying to find a tiny village in the middle of the countryside. As he drives down a narrow, winding lane he notices a farmer working in the fields nearby. He pulls over and stops to ask directions. 'How do I get to Little Windlesham village?' The farmer pauses, stroking his grey-stubbled chin thoughtfully, and then says: 'Well, if I were goin' there, I wouldn't start from here!'

Introduction

- **Where are you?**

- **Where are you going?**

- **What are you trying to do?**

- **Why are you doing it?**

- **Who is it for?**

Design process can be like this. It is important to know where you are, where you are going and how you are going to get there. What are you trying to achieve? Who needs to do it? Why? It is easy to lose yourself in iterative loops of design development and become dazed by details and endless refinement, as well as the many conflicting requirements and demands from many different people. Meanwhile you have missed the deadlines and the cost has gone through the roof!

An understanding of the structure of the design process is needed to give you the context within which you are working and, depending on the status of the stage, then the objectives that you are trying to achieve will change – and so will the requirements of your role. Just imagine – if you are undertaking a major journey, you will make plans, understand the landmarks and milestones on the route, and plan for any eventualities (as we all do, of course!) The design process is no different – it is a question of keeping in mind the route map for your journey with design on your project, understanding the twists and turns, key points and landmarks along the way, in order to reach your ultimate objective.

In contrast to the quote from Einstein, we are trying to make the DM process and success repeatable, while improving our performance each time we repeat the cycle – through review, learning and feedback.

The Design Manager's Handbook, First Edition. John Eynon.
© 2013 The Chartered Institute of Building. Published 2013 by Blackwell Publishing Ltd.

Get some 'TCQ' – Time, Cost, Quality in DM

The backbone to our understanding of DM is the mantra of time, cost and quality.

Using the analogy of planning a journey, we need to know:

Our journey with the farmer	*(Design Management, the process, activity)*
Our destination	*(Project definition – Quality, the end product)*
The legs of the journey	*(Stages or phases of the process)*
Milestones or landmarks on the way	*(Deliverables or outputs at completion of each stage)*
Time for the journey	*(Programme – Time)*
Things we need	*(Resources, people, organisations, technical)*
How much we need to spend	*(Cost, Budget plan)*

Just as our imaginary journey needs to be planned, the DM project process is no different. We need to break the process down into stages and understand the context and requirements of each.

Time, cost, quality, with the additions of safety and environment (including whole-life costs and management), represent the fundamentals of measurement for project performance.

In Section 3 TCQ is developed into a framework for Design Management for a project – the 'DMTCQ'. These are the aspects and activities that need to happen for DM to be successful and, if you're reviewing a process, then checking that these activities are happening correctly. The framework could also be used to signpost the rescue of a process if things are not progressing as they should.

There is also another strand running all the way through this, which is Value. Value Management and Innovation are discussed later in Section 6.

Our customers require their projects delivered on time, to their budget – providing value for money – and meeting their briefing requirements. As a by-product the project needs to be compliant with all statutory requirements, delivered without injury to anyone involved, and environmentally responsible in terms of resource usage and carbon emissions. Also it must be operationally economic and flexible enough to respond to future challenges and demands – particularly with respect to potential changes of use, adapting to increasing effects of climate change and also possible future retro-fits with new technologies.

Any project delivery process, including the Design Management aspects, needs to achieve at least these headline principles.

Defining the landscape

So let's begin with . . .

There are several project process models in the UK, each having a different slant on process and objectives. Before we consider them in turn, let's first consider a simpler view of the overall process, to provide an overall framework for our thinking.

In this sub-section, we will consider:

- A DM Definition

- The four stage process

- Typical role profiles for the Design Manager and the team organisation.

A definition of Design Management – The Activity

Design Management includes the management of all project-related design activities, people, processes and resources:

- Enabling the effective flow and production of design information –

- Contributing to achieving the successful delivery of the completed project, on time, on budget and in fulfillment of the Customer's requirements on quality and function in a sustainable manner

- Delivering value through integration, planning, co-ordination, reduction of risk and innovation

- Achieved through collaborative and integrated working and value management processes.

CIOB DM Working Group

Supplementary definitions

Activity (or Operation) – Actions by one or more businesses or organisations that enable another to commence, and produce a deliverable at completion, possibly completing at a programme milestone. The output might be information or a manufactured or built product.

Brief – Statement of client's requirements, forming initial basis for feasibility studies and design concepts. Can be expanded, clarified as the scheme develops: a living document, often a basis for Employer's Requirements together with other design information. BS7000 Part 4 and the RIBA Plan of Work describe the evolution of the briefing document at various project stages.

Builder's Work (in connection) – drawings show required builder's work in connection with other work, e.g. for mechanical and electrical services, indicating details of penetrations through columns, walls or floor slabs, or cast-in slots for fixings for cladding, etc.

Commercial management – management of project financial issues, contract.

Construction management – management of construction-related issues, logistics, safety, supply chain, programme, delivery).

Contractor – usually the main or principal contractor.

Customer – client, employer.

Deliverable(s) – The end product of an activity, or series of activities; might be produced at a milestone or on completion of a stage or phase; probably drawings, specifications, details, schedules or reports of some kind.

Designer – architect, technologist, structural engineer, civil engineer, MEP services designer, and other designers; specialist consultants who have an input to design.

DM – Design Management, the activity of *managing* the design process. Designing (*creating*) the 'design' is obviously related, but is a distinctly different activity. Designing is *distinct* from managing the process. Some DMers could do with learning this basic lesson.

DMer – Design Manager. This could be one person in the role of DM, or could be a team of people doing a composite team DM role, or could be another person such as a project manager, architect or bid manager doing DM as a subset of their skill set for their project role. The DMer could be part of the contractor's team, might be in the designer's team, could be in the customer's organisation, or could be a standalone consultant. Also could be office- or site-based.

Drawings – come in various types and are known by various titles. *General Arrangements* (GAs): overall key layouts, perhaps up to 1/100, 1/50 *Detail drawings*, *components and assembly drawings* might be 1/10, 1/5, or full-size to show the fully detailed construction. How the drawings are formatted and organised will depend upon the project type and size.

Milestone – a programming term that defines a significant point in the process: the completion of an activity, a series of activities, or a process stage. This point could result in a number of deliverables, or possibly be a hold point for a review or sub-process, or it could act as a trigger point for other activities or operations.

Project – a building, a built environment asset, a piece of infrastructure, or a constructed/manufactured/installed solution that has been designed. It could range from a house, to an office, a nuclear power station, a road, an airport or a development comprising several sub-projects, buildings and infrastructure. The term 'vertical' or 'linear' is gaining prominence. Vertical refers to a building or structure, while linear is more infrastructure-related.

Project management – management of the overall project, including everything related; responsible for the overall project outcome and its successful completion.

Schedules – for example for doors, windows, cladding, ironmongery, sanitary ware, etc. They could be a consultant's own version/format, but usually are now produced by the relevant sub-contractor/supplier as part of their tendering or quotation process.

Shop drawings –another term for the sub-contractor's design, manufacture and assembly drawings.

Specialist contractor or supplier – part of the project supply chain, also called works package contractor, or subcontractor to the main contractor. Sometimes they have a design responsibility, such as for steelwork connections, cladding, curtain walling, MEP services detailed design, etc.

Specifications – a specification is a statement of requirements, including technical standards to be complied with, e.g. workmanship standards, required quality and any other relevant information, which defines the designer's requirements for the specified element, component or section of construction. The document could be a consultant's own version, or may use the NBS format, the National Building Specification (www.thenbs.com/). The specification might be *performance*-based, meaning that a standard to be attained or designed to is described and usually combined with design-intent drawings, which then provide a starting point for subcontractor design development. This would be the approach on a CDP package (Contractor's Design Portion subcontract design-and-install package). Alternatively the specification could be fully prescriptive, setting out exactly what is to be provided and constructed, including the products, references, types, etc.

Stage Reports – it is now accepted practice for the Design Team to produce an end-of-stage report, which summarises the design progress to date, with individual reports by design discipline together with any specialist inputs (BREEAM, acoustics, fire, etc.), and highlighting any outstanding issues for approval, resolution or decision by the customer or the team. This is an extremely useful way of drawing a project stage to a close, and provides an ideal time for review of the progress and the team performance, before proceeding to the next stage.

Definition commentary

DM cannot function in isolation – in the context of overall Project Management, DM must work hand in hand cooperatively with Commercial Management and Construction Management at all times, to achieve the mantra of time, cost and quality.

DM is intrinsically connected – connectedness to all aspects of the project is the lifeblood of DM. Naturally this includes attention to aspects such as safety, sustainability and buildability. Connectedness and collaboration are how DM works; effective DM cannot function without them.

DM is a hugely diverse field and exists in many industries and technologies – not only construction. Therefore it is unlikely that any definition or framework will fit every particular situation perfectly. You will need to apply the principles to the peculiarities of your own situation.

In the case of a consultant DMer providing a DM service to a customer or contractor, then the same rules apply that we will discuss later in this section. Any contract of services must answer the Who–What–When questions, must be agreed up front, and must be absolutely crystal clear about the following questions:

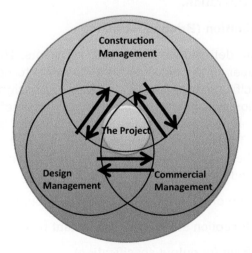

Figure 2.1 The Perfect DM project – all disciplines working together holistically for the good of the Customer and the Project – Connect, Integrate, Communicate, Innovate, Collaborate – CICIC principles – DM is intrinsically connected – it has to be!

- What are the project objectives?
- What is the project budget?
- What services are being provided?
- Who is doing them?
- Who is needed for support or input?
- Who needs to be consulted and informed?
- What needs to be done?
- What are the outputs, deliverables?
- What are the timescale(s)?
- How much will it cost?

Ideally, as the DM process continues through the project, we should be steadily building on the previous stages, using and adding to the knowledge that has previously been gained and maintaining the continuity of flow. Frequently, however, this is not the case: whole teams and individual participants can be changed, resulting in knowledge being lost. The process unravels as previous decisions are reviewed and changed. Discontinuity and disconnection are the enemies of project success.

Standing between design and construction, or between pre-construction and delivery, can be an uncomfortable place. There are tensions not only within these two camps, but also between them. The DMer needs to be able to manage these situations and respond accordingly, while keeping the whole process on track.

Four stage process

Generically speaking there are four main stages in project process (as shown in Figure 1.1):

- **Project Definition**
- **Project Procurement**
- **Project Delivery**
- **Project Operation.**

Project Definition (Briefing/Scoping)

This is about defining what the customer wants. The brief might be simply one sheet of paper, or might range all the way up to concept drawings, room data sheets, specifications and so on produced by a design team. Whatever the amount of information available at this point, that is what defines the project design brief, although it is common practice for the brief to be developed as the design progresses. The brief develops through dialogue with the design process. However, for clarity there needs to be a definitive brief to enable evaluation of the design when required.

The RIBA Plan of Work recognises brief development in this way:

- Brief at Inception stage, initial statement by the customer
- Design Brief (or output specification)
- Project Brief, subsequently developed from previous briefs.

(RIBA Plan of Work)

Project Procurement (Pre-construction)

This stage is about developing the project definition information into the project design and related information sufficiently in order to be able to tender the project and to appoint the main contractor, i.e. to procure the delivery of the scheme.

The design information produced is the means to this end, i.e. to enable the construction of the project.

Design status at the point of tender or bid is extremely variable, ranging from, say, RIBA Stage D to Stage F1. Procurement routes vary from single- and two-stage design and build, various forms of traditional contract, and also partnering-style arrangements. Integrated Project Delivery (IPD) and Integrated Design Construct (IDC) assemble the main team, including designers, contractor and key suppliers, at the start of the briefing/definition stage.

The information issued for tender will define the Employer's Requirements. The contractor's response, particularly where a design response on a design-and-build-style contract is required, form the Contractor' Proposals (i.e. this is an offer defining time, cost and quality for the project).

Project Delivery (Construction)

Through the tender process in the previous stage, the project envelope has been determined in terms of time, cost and quality. What is required next is the information needed to physically construct the project. The DM challenge here is that the design must progress from whatever stage had been reached at tender into full construction and installation information. This will involve integrating the work of the design consultants, and specialist subcontracted designers, while probably also having to start on site as quickly as possible.

Key in this process is an Integrated Design/Procure/Construct programme, which brings together the activities and deliverables of everyone involved: designers, specialist consultants, contractors, supply chain and subcontractors. If there is a substantial overlap between the design development period and the site operations (and there usually is), then in developing the design programme the DMer will need to address how information that is sufficiently resolved and co-ordinated is released in time to enable procurement of packages and the construction operations to continue smoothly.

Project Operation (Building in use/Facilities management)

The period before and after Practical Completion is critical. This includes completion of snagging, quality issues, commissioning of building systems and training for the building end users. At Practical Completion, handover of the Health and Safety File and the Operations and Maintenance Manuals is required, plus completion of any training for the customer, their team and end users.

Aftercare and post-PC support has now become much more important. All that good work in developing the design, building it to the required standards and delivering on time can be thrown away in a handover process fraught with defects and problems. Happy clients mean repeat business, so ensuring that this crucial part of the project delivery process goes smoothly can provide substantial benefits.

BSRIA have developed the '**Soft Landings**' process to deal exactly with these issues. For a systematic approach to building commissioning, handover issues, post-completion methodology and post-occupancy evaluation, then follow this link for further information.

BSRIA Soft Landings: www.bsria.co.uk/services/design/soft-landings

Roles

The role of the Design Manager will change depending on the stage of the project. On our DM working group we have looked at different roles, particularly pre-construction and site delivery:

Delivery Design Manager – is frequently site-based and is usually a member of the main contractor's team. At this point the delivery parameters for the project have usually been set, i.e. programme time, contract cost, quality and the content of the project to be completed.

The role consists of ensuring that consultant designers develop the design to a sufficient level of detail and also that the integration of the work of the specialist subcontractors takes place, ensuring that the correct information is being released in a timely fashion to suit the construction operations.

Co-ordination of the design information, in terms of integrating design inputs, e.g. structures, MEP services with the architectural design, is usually performed by the lead designer, in most cases the architect. On projects where there are substantial MEP services or structure/civils content, then the lead designer would normally be from the relevant discipline.

Pre-construction Design Manager – could be from the main contractor or possibly from the lead designer's team, managing the process of taking the project design brief and developing the design and associated information to a point where the project can be tendered under the procurement route being used. Included in this process will be managing the input of other specialists, e.g. acoustics, rights of light, sustainability, etc., plus possible early involvement of specialist subcontractors to clarify design, cost and buildability issues. Value management and cost management are critical activities together with the design processes themselves, to ensure that the developing design meets the brief but also meets budgetary constraints.

Strategic Design Manager – Possibly a consultant to the customer? The SDM deals with the very earliest stages of the project, assisting with the development of the business case and preliminary objectives for the project and defining the project brief sufficiently to enable the procurement stage to commence.

After Practical Completion/handover, the SDM deals with the whole life issues of the project, including Post-Occupancy Evaluation, FM-related and operational issues. Resource and energy use and carbon emissions will become increasingly important. Should the need arise for further refurbishment, alterations, extension or other development of the project, then the SDM will work with the customer in developing the next cycle of the process, looking at options, appointing specialists and so on.

Profiles – Included here are two generic profiles for the Pre-construction and Delivery Design Managers. In our DM working group we have looked at several examples of profiles and collated our thoughts. These profiles are not so much a prescriptive outline of the roles, but more an agenda. In our discussions we recognised that Pre-construction (characterised by *opening up*) is a different world from site delivery (characterised by *closing down*) –meaning opening up or closing

Pre-construction Design Manager – 'Opening Up'

Pre-constructionfocus
Usually off-site

**Overview
– Optioneering, compliance and procurement**

Once the project direction has been established and the brief formulated, then the design is developed for submission for planning approval and further developed for Building Regulations approval and fortender/negotiation with the main contractor. There may well be the need for further value-management exercises to be held to ensure the project is within budget, butt his will usually be within the now-established project parameters. Sometimes it becomes necessary to go back in the process because requirements change, budgets are exceeded or other external factors come into play. Other activities could include early supply-chain input, review of key aspects and risk mitigation. These are activities that need to happen. Time and expertise willing, these activities could be carried out by a Project Manager or Bid Manager, as part of their skillset, for example. Interms of site delivery, the DM function could be fulfilled by a Package or Site Manager responsible for the construction of the design elements involved. Cladding, MEP service, structures would be good examples of where a suitably able and skilled Package Manager would deal with the package from design through to final completion as part of their responsibility. This could equally be applied to any aspect of the construction, once the overall parameters have been set.

Forming the time cost quality delivery/realisation 'envelope' for the project/asset(TCQ).

Continuity and responsibility to ensure that knowledge is transferred and retained across stages, and not lost.

Key drivers

Understand the customer's value drivers
Brief management – clear brief, ERs, CPs, critical to avoid future ambiguity
Determining scopes
Determining design team
Determining fee proposals
Appointment documents
Key subcontract design packages
Risks, and opportunities and possibilities
Opening up ideas/options and closing down for procurement
Preparation for effective construction process
Tendering function
Manage Customer expectations
Safety - CDM

Activities

Leadership – leading the team
Reviewing tender information
Buildability and alternative innovative solutions
Affordable
Sustainable
Understanding the vision, brief, monitor delivery of customer's requirements/brief
Meets customers
Ideas
Needs involvement post contract
Manage value, implement VM process
Manage people (designers)
Scopes of services
Resourcing
Contractual relations
Contribute to submissions
Review performance of designers, KPIs
Stage-completion reviews
Lessons learned
Areas for improvement

Delivery Design Manager – 'Closing Down'

Construction delivery focus
Probably site - based

**Overview
– Co- ordination and delivery**

Once the scheme is defined and approvals obtained, the project moves into delivery mode. The objective at this stage is to feed the construction machine the information it needs to complete the project to programme and budget. This involves integration and co-ordination of all the stakeholders, including designers and supply chain. Programming, cost, buildability, compliance issues, all come into play, and at this stage the site-based Design Manager has to deal with all these issues.

These are activities that need to happen. Time and expertise willing, the activities could be carried out by a Project Manager or Bid Manager, as part of their skill set, for example. In terms of site delivery, the DM function could be fulfilled by a Package or Site Manager responsible for the construction of the design elements involved. Cladding, MEP service, structures would be good examples of where a suitably able and skilled Package Manager would deal with the package from design through to final completion as part of their responsibility. This could equally be applied to any aspect of the construction, once the overall parameters have been set.

Delivering within the project 'envelope' (TCQ).

Continuity and responsibility to ensure that knowledge is transferred and retained across stages, and not lost.

Key drivers

Risk, practicalities
Closing down
Control of detailed design (under D+B or CDP)
Radar for problems and facilitating solutions
Design to cost
Integrating the delivery team
Technical query resolution

How does design optimise production on site?
Design for production, and performance

Activities

Delivery
Meets suppliers
Needs involvement pre-contract
Programme and deliverables management

Manage value, VM not just VE cost cutting
Stuff that nobody else has bothered to sort
Relational contracting
Design team meetings/record of
Integrator
Engaging whole team

Outputs

Detailed design information at the right cost at the right time for construction
Design reports
Feedback on performance
KPIs to measure team's performance
Compliance, statutory, regulations
Planning, Listed, Building Regulations, BREEAM, Customer

Figure 2.2 Generic DM Role Profiles.

2 Process

Pre-construction Design Manager – 'Opening Up'

Outputs

Responsible to determine scope
Project Brief
CPs
Winning scheme
Designers' agreements
Design scope of works
Programme for design

Skills

Managing (people and process)
Client facing/selling/presentation
Empathy
Agility
Engage right people
Set up and lead the 'design team'
Ethos and spirit
Good communicator
Negotiator

Experience

2 years' relevant experience
Unique position to influence and advise customers and designers
Understand and explain how customer's or contractor's process affect design
Worked as Delivery DM on some projects, minimum 2 years

Qualifications

Minimum degree level or equivalent (work experience)
Formal qualification and membership of professional institute

Person

Have confidence to ask the questions, to think the unthinkable – to get beyond 'we've always
done it this way – why?'
Design-orientated, some appreciation not necessarily from a design background
Comfortable with ambiguity
Taking something uncertain to a level of certainty
Results-focused
Team leader, collaborator, integrator

Note – understand impact at operational level (delivery process, functions, problems) before
moving to front-end work/pre-construction
Avoid mismatch between 'what we do' and 'what customers want'
Understand customer business drivers – adding value beyond the box is becoming more
important in a competitive industry

Delivery Design Manager – 'Closing Down'

Skills

Client facing/managing
Methodical
Process management
Cope with conflict and manage it effectively
Timely review and comments on deliverables
Good communicator
Good grasp /appreciation of design
Ability to say 'No'
Ask 'Why'
Flexible/agile – manage change, opening up in a closing-down environment – life happens
Managing meetings
Technical knowledge

Experience

2 years' relevant experience
Worked alongside Delivery DM on some projects, minimum 2 years before advancing to own
projects. Previous site experience useful.

Qualifications

Minimum degree level or equivalent (work experience)
Formal qualification and membership of professional institute

Person

Delivery-oriented
Uncomfortable with ambiguity
Managing something certain
Results-focused

Ability to rise above detail sometimes, to take a longer view on project value, sustainability,
impact of decisions, customer value

Figure 2.2 (*Continued*)

down options and ideas. In pre-construction we want to open up ideas and explore options to obtain the best solution and also work towards developing the project envelope; but in delivery we want to close down and focus in on the solution and deliver that as efficiently and effectively as possible.

Our thinking went as far to conclude that these roles actually need different kinds of people, perhaps even down to different mindsets, profiles and even personality types. So these indicative profiles represent a jumping-off point for your own thinking and consideration, in your own business and project situations. From these you could develop them to the next level in terms of business and process-specific activities.

Sitting on the fence between design and construction can be an uncomfortable place to be. There are many pressures upon the role. At various times in the past,

I have been criticised for being too much on the side of either the designers or the contractor, depending on who was speaking at the time. While the joys of the role in bringing the project together and seeing the complete picture emerge cannot be overstated, the associated challenges should not be underestimated by the budding DMer.

The ability to handle many inputs and demands, sometimes conflicting, is a key aspect of the role, at every stage of the process.

Titles

Job titles for the DM role are extremely variable, but generally the main core activities of the role are relatively very similar across the industry. Some contractors have refrained from using the word 'co-ordinator' in their role titles, as it implies a co-ordination responsibility. The lead designer/lead consultant on the project will usually be responsible for co-ordination and integration of the design disciplines and those tasks will be included in their duties in their appointment documents, which you need to check.

Before we launch into other processes in a little more detail, let's consider a few other issues.

Starting and finishing

Whatever process you use, the aspects of starting and finishing stages are critical activities in themselves. These moments need and deserve a lot of attention.

In starting a stage you have to be clear about what is to be achieved: all the 'Who, What, When, Why, How' questions etc. need to be answered.

How will you know that you have completed the stage as you have defined it? What are the outputs?

Finishing a stage is equally as important:

- Have you achieved all that you planned to do?

- Are any activities incomplete?

- How will these be managed or carried through to the next stage?

- Are the deliverables and outputs as you need them to be?

- Are there any issues outstanding that require resolution before moving to the next stage?

- Are there any lessons to learn?

- Does the overall design process plan need adjustment?

- Do you need more time? Resources? Specialists?

In our rush to comply with our programmes and deadlines, it is worth spending a little time to coolly review all the activity from time to time. Sometimes a little more thought, and a little less action, will pay dividends. Peer review can be really helpful here. A fresh pair of eyes looking over your shoulder can be a real help. (This is discussed further in Section 4: Tools.)

2 Process

> **Finishing well**
>
> It's not enough to finish the checklist, to hurriedly do the last three steps and declare victory.
>
> In fact, the last coat of polish and the unhurried delivery of worthwhile work is valued out of all proportion to the total amount of effort you put into the project.
>
> It doesn't matter how many designers, supply chains, workers, materials and factories were involved – if the box is improperly sealed, that's how you will be judged.
>
> *Seth Godin's Blog – Seth Godin (http://sethgodin.typepad.com/)*

Finish all the stages, tasks, deliverables properly; do not just stop. Finishing and stopping are completely different things!

The overall project context

In this section we consider the context of the design and construction process against the backdrop of the project lifetime.

A typical building project, once completed, currently could last for up to 100 years and in some cases longer.

A large proportion of the housing stock in the UK dates from Victorian times, and is still in serviceable condition, proving wonderfully flexible and responsive to succeeding generations.

So from the time of that first glint in a customer's eye indicating that they want a new building, to when that building has seen out its operational life and is due for demolition and recycling, could quite easily be 100 years or more.

Where is the value?

Looking at the project process from the outset, it could take a few years to establish a business case, define the need, get a design team together, obtain planning permission and put the scheme out to tender. Construction might take one to two years or more, depending on the size of the project and its complexity.

As an industry we expend huge amounts of effort, resources and money on the design and construction processes – these represent perhaps 1–2 per cent of the lifetime of the building. A Constructing Excellence study several years ago revealed that for a typical building project, the relative costs for construction, maintenance and operation were in the ratio 1:5:200. So the operational expenditure of a building (the Opex) far outweighs the Capital Expenditure (Capex).

So concentrating purely on the construction phase, how much additional value can we realistically create? Our surveying friends might achieve some buying gain by letting some packages at a cheaper price. But this might be at the cost of quality or the services provided, which could lead to further problems later.

Analysis of site logistics and constructability might yield some additional margin, but we are looking at perhaps a few percentage points on the construction cost,

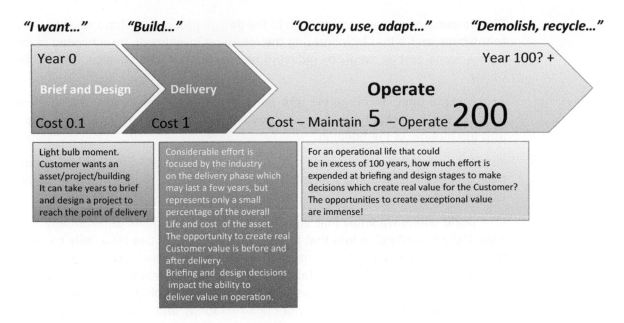

"*I want...*" "*Build...*" "*Occupy, use, adapt...*" "*Demolish, recycle...*"

Year 0

Year 100? +

Brief and Design

Delivery

Operate

Cost 0.1

Cost 1

Cost – Maintain 5 – Operate 200

Light bulb moment. Customer wants an asset/project/building It can take years to brief and design a project to reach the point of delivery

Considerable effort is focused by the industry on the delivery phase which may last a few years, but represents only a small percentage of the overall Life and cost of the asset. The opportunity to create real Customer value is before and after delivery. Briefing and design decisions impact the ability to deliver value in operation.

For an operational life that could be in excess of 100 years, how much effort is expended at briefing and design stages to make decisions which create real value for the Customer? The opportunities to create exceptional value are immense!

Figure 2.3 Project Timeline.

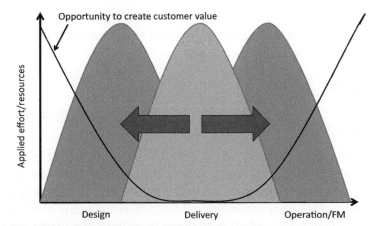

"The Three Humps" – Most of our effort is focused on the procurement and delivery stages.
The opportunities for creating Customer value lie before and after the delivery phase – we need to change the focus of our efforts!

Figure 2.4 The Three Humps

which itself represents at the most a few per cent of the lifetime cost of the building for the customer.

Isn't this a bit of a futile effort? Couldn't our effort be applied elsewhere for more substantial benefits – both for our customers and ourselves?

Leveraging value

I believe we need to be looking both earlier and later in the project timeline. The opportunity to create value increases the earlier we intervene in the process, as does the value created. Integrated models of procurement such as IPD or IDC bring the contractor into the project from the start. This enables the right decisions to be taken that affect the delivery phase at the start, rather than having to revisit them later in the process.

Simple decisions taken early enough in the design process can have huge benefit to the customer, either by increasing the return on investment, by achieving more for the same cost or simply by reducing risk from a particular design, method or option.

At the same time, decisions we take early in the design process can also have far-reaching implications that go way beyond Practical Completion, many years into the lifetime of the building.

Just suppose that we take design decisions that slightly reduce energy consumption, improve the building's adaptability to change and reduce its reliance on external energy sources.

In a world where we know that energy costs are going to rise substantially and potentially the infrastructure that provides the energy we use on a daily basis is going to be overloaded, decisions such as the above could save the customer several times the building's capital expenditure over its lifetime, as well as saving carbon and energy in the process. We could equally look at maintenance and replacement cycles, consideration of constructability and logistics and procurement methods, to ensure that the whole process provides the customer with an efficient, optimised building solution that will continue to deliver benefits and value long after the design and construction teams have moved on to their next site.

We will consider value management and creation together with innovation later in the Handbook, in Section 6.

Integrating the team

If we take the above observations on board, then this leads us to a few conclusions.

Firstly, even the simplest projects are pretty complex technically these days. They require specialist input, which is beyond the expertise of the main consultant designers and the main contractor (the latter essentially being a management contractor – very few major contractors employ any direct labour; virtually everything is subcontracted, even down to welfare and waste management).

Secondly, as we have seen, the opportunity to create value increases the earlier in the process it can be applied; therefore we need to apply our best intellectual effort early in the design process to ensure that the best solutions are adopted and the cost of change is minimised. Once ideas are more fixed, to change course incurs cost penalties, which can be substantial due to design changes, and in the case of built work may be prohibitive.

Therefore, early in the design process all the main specialists should be involved. I include the main contractor and the key supply-chain members. Frequently customers and their teams are now inviting contractors to give pre-tender advice on logistics, programme, constructability and some supply-chain input where appropriate.

Early intervention and integration will produce value for the customer and produce a successful project environment for the team. This does raise issues around how projects are procured, but there are plenty of examples of Integrated Project Delivery and projects using this approach on Partnering Contracts.

If the appropriate intellectual input and effort are made at the right time, up front, then the later stages of design procurement and construction can be made much more straightforward in terms of design and delivery processes.

Process models

There are a number of project process models in use across the industry. They all have their reasons for existing, whether it is a particular discipline's or industry group's view on life, or a particular focus on certain aspects of the process.

In this section we will consider the following:

- BS7000: Part4: 2012
- BS1192: 2007
- CIC Scope of Services
- The RIBA Plan of Work
- Salford Process Protocol
- OGC Gateway™ Process
- Avanti
- DM and CDM
- DM and Cost Management
- The Last Planner.

The RIBA Plan of Work is perhaps the most widely known and accepted in the UK building industry. Nevertheless it is a design-centric process, and although it covers the whole lifetime of the project, it is really about design process. It has been opened up to a more multi-disciplinary project view, but naturally still represents an architect's view on the process. Amendments have considered different procurement strategies, a green design issues overlay and also a BIM overlay. The latter is currently under review at the time of writing.

Process Protocol, on the other hand is much more a project process from a business perspective, which happens to cover design as well. It is a flexible process, which offers the shades of detail that include integrated team working at all stages. A particularly useful aspect is stage gateways and reviews

If you work on schemes in the public sector, then you will need to understand the OGC Gateway™ Process. It is supported by an excellent website and various publications to help the project team understand what is reviewed at each stage and how to achieve excellence and best value.

The CIC stages and scopes have gained broad industry acceptance and are currently being used by the various industry BIM working groups to align processes with UK government strategy.

Before we move on, a word of caution about processes. I think we are fooled into complacency by our process charts and checklists. It all looks so convincing, doesn't it? – design stages, deliverables, outputs, reports. Our teams excel at producing information by the lorry-load. However, do we ever complete anything properly? Do design stages finish completely with everything resolved? Are the stages' activities completed, or do they just stop because time has run out? Whatever is left unfinished is then dragged behind the project team as an accumulating bag of risks, which will come back to make their presence felt in the delivery stage.

Just because you have produced a programme or process chart with a comprehensive set of checklists does not mean all the problems have been solved. People are not machines or like chess pieces. Although the chart tells them they are supposed to do it, that does not mean they will!

BIM, process and the future

During 2011 in the UK BIM suddenly seems to have finally captured people's interest, and achieved weekly coverage in the building press – no doubt fuelled by Paul Morrell's pronouncements regarding BIM and access to public-sector work! BIM will be a requirement for all public-sector projects from 2016. At the same time the government is looking for 20 per cent savings in the cost of delivering construction projects.

It depends who you talk to – but is this Building Information Modelling or Building Information *Management*? BIM comes in several flavours and also in levels 1, 2, and 3, as well as with names like lonely BIM, selfish BIM, iBIM, ndBIM, BIM(M) and so on.

Certainly BIM offers the potential of a different way of working and encourages integration of the whole team to collaborate and work together more effectively.

BS1192, and the rewrite of BS7000 Part 4, provide a framework for teams to work in a BIM environment, as do several publications by the American Institute of Architects – which, while relating to USA-style procurement, give a lot of clues as to how to frame a BIM project process.

It seems to me that BIM changes or has the potential to change everything in our industry, and I don't think that's an overstatement. It touches all stages of the project and provides a technology base for us to achieve what previous generations have only dreamed about. From initial concept, through detailed design, computer-driven manufacture and construction installation to facilities management – BIM provides an integrated information-management solution, crossing the boundaries between consultants, supply chain and manufacturers and constructors. The UK has been slow on the uptake, but that is changing, and I believe will create a watershed in the building industry much greater than that created by CAD in the 1980s. It will impact every aspect of design, procurement and construction and how projects are commissioned and delivered, including the function of Design Management and the role of Design Managers in general.

So for the purposes of this section I am looking at DM processes pretty well as we know them now, i.e. using a pre-BIM or non-BIM approach. It will be several years before all the UK industry is fully BIM-integrated, so DMers will have to mix and match their processes, tools and approaches until we all get there – wherever 'there' is.

We will consider BIM, and how it could affect project process and what tools are out there, in the final section of the Handbook. It will be the case that for quite some time different systems will be working in parallel, depending on the capabilities and resources of the businesses, supply chain and individuals involved.

We should not be distracted by the technology, awe-inspiring as it is. The project team, the people, the process, the inputs, outputs, information flows, etc. all still need to be managed. While the processes and technology may be quite different from what we are used to currently, the right actions need to happen at the right time, just as now, for projects to ultimately be successful. DM as a role and a critical activity will continue to exist. It is perhaps that the role name or lapel badge will change and the playing field will be a little different.

B/555 Roadmap (August 2012 Update)

Design, Construction & Operational Data & Process Management for the Built Environment

Introduction

The purpose of this updated "Roadmap" is to document and describe the activities of the BSI B/555 committee (Construction design, modelling and data exchange) in the immediate past, current and future in support of delivering clear guidance to the UK industry dedicated to providing and operating built assets. It also supports the vision and mission statement of the committee in the reduction of whole life cost, risk, carbon and the timely delivery of buildings and infrastructure projects. The 2012 Update also incorporates the activities in support of the 2011 HMG BIM Strategy.

To illustrate the process a maturity model has been devised to ensure clear articulation of the standards and guidance notes, their relationship to each other and how they can be applied to projects and contracts in industry.

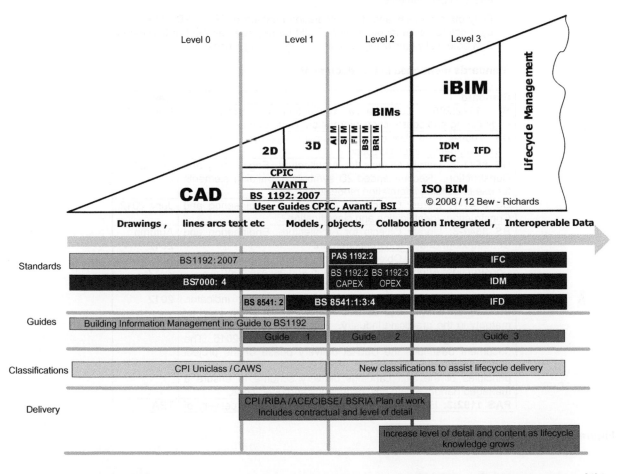

Figure 2.5 BSI B/555 BIM Road Map. Extracts reproduced by permission of BSI Group. The latest version of this document can be obtained free of charge from the BSI online shop: http://shop.bsigroup.com/bim

To simplify the description of technologies and ways of working, the concept of maturity "Levels" has been defined. The purpose of the maturity levels is to categorise types of technical and collaborative working to enable a concise description and understanding of the processes, tools and techniques to be used, thus allowing simple referencing as to where various documents should be applied.

Maturity Level Definitions

0. Unmanaged CAD probably 2D, with paper (or electronic paper) as the most likely data exchange mechanism.

1. Managed CAD in 2 or 3D format using BS 1192:2007 with a collaboration tool providing a common data environment, possibly some standard data structures and formats. Commercial data managed by standalone finance and cost management packages with no integration.

2. Managed 3D environment held in separate discipline "BIM" tools with attached data. Commercial data managed by an ERP. Integration on the basis of proprietary interfaces or bespoke middleware could be regarded as "pBIM" (proprietary). The approach may utilise 4D Programme data and 5D cost elements.

3. Fully open process and data integration enabled by IFC / IFD. Managed by a collaborative model server. Could be regarded as iBIM or integrated BIM potentially employing concurrent engineering processes.

Standards mentioned in this document

Standard	Date
BS 1192:2007 Collaborative production of architectural, engineering and construction information. Code of practice	Dec 2007
BS 7000-4:1996 Design management systems. Guide to managing design in construction	Early 2013
BS 8541-2: Library Objects for Architecture, Engineering and Construction: Recommended 2D symbols of building elements for use in building information modelling.	Sept 2011
BS 8541-1: Library Objects for Architecture, Engineering and Construction: Identification and classification	June 2012
BS 8541-3: Library Objects for Architecture, Engineering and Construction: Shape and measurement	Autumn 2012
BS 8541-4: Library Objects for Architecture, Engineering and Construction: Attributes for specification and assessment	Autumn 2012
PAS 1192:2: Early Adopter document to enable the delivery of HMG BIM strategy projects to Level 2 maturity indicator. Created as a PAS due to time constraints and low level of maturity in the practicing industry. Expected delivery late 2012, with full upgrade to British Standard level before 2015. The document describes the Capital delivery phase of the project during design and construction. The document incorporates the principles of the Soft Landings delivery scheme to ensure a managed handover into post occupancy and operations.	Autumn 2012
PAS 1192:3: Early Adopter document to enable the delivery of	TBA

Figure 2.5 (*Continued*)

HMG BIM strategy projects to Level 2 maturity indicator. Created as a PAS due to time constraints and low level of maturity in the practicing industry. Exact dates to be advised due to budget constraints.	

The application of standards is dependent on many often poorly understood or articulated factors. The maturity model is used to identify where standards and associated tools and guides are applied to develop a coherent solution to inform the delivery process.

The B/555 Roadmap deliveries are related to the appropriate "Level" to aid clarity of application within the operational and delivery market. July 2011 marked the formal release of the UK Government BIM Strategy. The effect of this has been to set a focused timescale for the adoption of technologies and process to deliver significant cost and carbon performance improvement across the industry. Key to the strategy is the need to deliver Level 2 BIM capabilities with combined model, drawing and COBie data deliveries to the client at key points throughout the delivery and handover process. To achieve this clear contractual and delivery guidance is being made available to the supply chain. The PAS 1192:2 and 3 form the first parts of this guidance. On successful completion of the Early Adopter programmes the documents and processes will be refined and converted into full British Standards.

Key Road Map Deliveries

Delivery 1 2010/12 - Object Libraries

Items indicated in green are existing documents available in the market today. BS 1192:2007 is a combined data and process standard and is equally applicable at level 0 and 1. It offers advice for the management of traditional CAD managed data delivery and works with both paper and electronic formats. CPI and Avanti have produced guidance to support implementation of BS 1192:2007 and BSI/CPI have jointly published a guide "Building Information Management - A Standard Framework and Guide to BS 1192" in September 2010.

New deliveries from B/555 are an update of the BS 7000:4:1996 to document the overall design coordination process and the planned management of the project data delivery synchronised across all participating disciplines.

Symbol definitions for the presentation of 2D information have not featured in the B/555 document family since the withdrawal of BS 1192:3:1987, this situation has been rectified with the release of BS 8541:2:2011. These symbols are predominantly for use at Level 1 and as such form a useful reference in the current market as well as a consistency through the maturity levels.

There has never been a consistent set of 3D libraries or definitions in the UK. This is a significant gap as 3D technologies are now commonly available in the market. BS 8541in its various sections will address this issue.

Figure 2.5 *(Continued)*

2 Process

BS 8541:1 Will introduce library objects, represented in appropriate formats for use at level 0 (blocks, cells) through to level 3 (IFC objects). The document will refer to BS1192:2&3 and object based principals for identification (naming) and grouping (layering and classifications), it will also include identification of source.

BS 8541:3 Will define 3D symbols in multiple levels of detail. This is essentially focussed on levels 1 to 2, to represent the analysed and designed output as the first level representation in a real world, they will include functional and geometric quantity measures (volume, projected area, plan area, effective length etc).

BS 8541:4 Will define properties and multiple levels of information; this will be essentially focussed on levels 2 to 3. The document will include
> Properties required for specification/selection
> Environmental, cost and social impacts (CEN/TC/350 Sustainability of Construction Works)

Delivery 2 2012/15 – Process and Data Management

Indicated in red are the new standards documents which build on BS 1192:2007 and enable us to make use of the various new technologies. It is clear that as new technologies and collaboration techniques come to market even more clear guidance needs to be made available. This guidance must be specific to its intended audience as the needs of clients, suppliers and users differ significantly. For this reason the documentation will be provided in two documents, the first focusing on the "Capital Delivery" phase and the second on "Operational Delivery" issues. Both will document both data and process management issues. Key issues dealt with will include:

> Process definitions
> Data management for data definitions used for
> o Production & operation
> o Libraries & specifications, properties and representations in various stages
> Generic Delivery Schedules identifying key deliverables at identified stages for all design, delivery and operational disciplines.

It is expected that existing classification and delivery schemes such as the RIBA stages etc will become compliant with these standards.

The two documents will be labelled as follows

> BS1192:2 Information management for the capital/delivery phase of construction using Building Information Modelling.
> BS1192:3 Operational Asset Management - Processes and data for the commissioning, handover, operation and occupation stages.

Figure 2.5 (*Continued*)

The two documents will deliberately overlap to ensure there is documentation covering the whole lifecycle from end to end. The definition of open data exchange between all stages including construction to operation will be defined. As mentioned above the early releases of these documents will be in the form of Publicly Accessible Standards (PAS) in recognition of the programme and adoption constraints driven by the HMG BIM strategy.

Delivery 3 2012/15 – Guidance Documents

Guidance documents are not seen as part of the remit of B/555 but as documents that will be delivered in partnership with the British Standards Institution. Example: Building Information Management – A Standard Framework and Guide to BS1192.

With such a complex subject clearly a significant level of clear supporting guidance will be necessary, to ensure consistence and quality, B/555 will coordinate the production of this material. The documents described below are indicated on the Maturity Model in blue.

Guide 1 Level 1

Building Information Management – A Standard Framework and Guide to BS1192 (2007) was published in 2010. It offers clear and detailed guidance as to the application of the standard in a pragmatic and clear form.

Guide 2 Level 2

Will offer guidance on the design, data management and the workflow processes to deliver the CAPEX & OPEX standard. The requirements and content as defined in the 'Delivery' documents to be produced by CPI/Avanti. These will contain the coordinated deliverable of each stake holder, architect (RIBA), structural (ACE) civil (ICE) and MEP (BSRIA) engineers against the RIBA Plan of Work Stages. For the infrastructure works we will include the railway (GRIP) stages.

Guide 3 Level 3

As maturity level 3 becomes a reality and technologies develop into web services and distribution of interoperable data sets a Level 3 Guide will be developed.

Figure 2.5 (*Continued*)

Appendix 9 of the BIS BIM Report illustrates the correlation between process model stages, and provides a useful comparison for reference. Further work on industry process alignment for BIM will be forthcoming from the various working parties.

We now turn to the processes.

BS 7000: Part 4

Design Management Systems, Part 4 – Guide to managing design in construction

This standard is currently under revision and is due for publication some time in 2013. It gives guidance on management of the construction design process at all levels, for all organisations and all types of construction project. The guidance covers management of design activities throughout the whole life cycle of a construction project. It covers subject matter such as the design team, the briefing process, project and process planning, programming, communications, cost control and the role of the DMer. It also provides guidance on design resource and design process management, design-team activities and post-occupancy.

This standard is fairly general and normative in nature, and provides a framework for design management on any project. It fits in under BIM levels 0 and 1 on the BSI Roadmap (see Figure 2.5) and may extend to level 2.

A common plea I have heard from Design Managers is for common processes across the industry. Certainly BS7000: 4: 2012 and BS 1192: 2007 provide that platform. While they may not provide all the answers, perhaps a first step in developing a common approach is to use the standards we already have?

The BIM working parties will progress the formation of aligned processes, particularly focused on the CIC and RIBA Plan of Work.

BS1192: 2007

Collaborative production of architectural, engineering and construction information – Code of Practice.

This was published to provide a framework for the development, organisation and management of production information. It is also applicable to a BIM project environment.

The BS considers roles and responsibilities, definitions, Common Data Environment, file conventions and specification, and provides appendices giving some templates and proformas.

It is an information-organising and -handling process. It relates to the work by CPIc, the Construction Project Information Committee, which has developed a number of BIM protocols referred to in the BS1192 Guide.

BS1192 is supported by the UKCG, CIAT, RIBS, CIBSE, RICS, ICE and RIBA. Therefore it is well worth getting to know this standard and its principles, as it is one of the few documents related to DM that has virtually industry-wide recognition and acceptance.

The Guide identifies several roles, including that of the 'Design Co-ordination Manager' (also known as the Design Manager on some contracts). The BS defines this role as providing a communication link between the design teams and constructors. It recognises that the DM is usually provided by the contractor and integrates the design deliverables of the professional designers, specialist designers and subcontractors against the construction programme, to ensure timely delivery.

Other defined roles include Task Team Manager, and Interface Manager. It could be argued that these are also part of the normal DM role (BS1192 Guide, page 13, paras 4.1, 4.3, 4.4). In BIM circles there is also the developing concept of the 'Project Integrator' or 'Data Manager', as fundamentally it is data/information that is managed over the asset life cycle – this data could be about design, O+M, FM, etc.

The BS1192 Guide not only recognises the DM role, but also that it is predominantly in the domain of the contractor. This returns us to the theme of the consumer contractor I touched upon earlier in the Introduction to the Handbook.

It is also worth noting that, as the Guide says, when working traditionally drawings are the deliverables. In a BIM environment, the model with all its data is the deliverable. In the 'traditional' DM process we manage outputs, i.e. drawings, schedules, etc.

The guide to BS1192 makes reference to the Avanti process, which is a collaboration approach to working in an ICT environment.

CIC Scope of Services 2007

The CIC has produced a series of integrated documents for scopes of services and consultant appointments.

These are based on project stages as follows:

Stage 1 – Preparation

Stage 2 – Concept

Stage 3 – Design Development

Stage 4 – Production Information

Stage 5 – Manufacture, Installation and Construction (MI&C) Information

Stage 6 – Post Practical Completion.

Stages 3, 4, 5 and 6 have both Design and Review components, which recognise the different services that, say, a Contractor's Design Team and an Employer's Review Team might carry out for compliance.

In my experience, where the design team has already been appointed prior to engaging the main contractor, the different consultant appointments are usually quite different. It is often difficult to check that there are no gaps between disciplines.

The beauty of the CIC suite of documents is that they are an integrated set of agreements and services for an integrated team. It is straightforward to cross-relate the different disciplines and understand the scope of the team as a holistic entity.

The CIC documents refer to the BSRIA's *A Design Framework for Building Services*. This is a suite of MEP Building Services-related documents, which uses the same stages and terminology and provides a further level of detail. See Tools in Section 4.

The RIBA Plan of Work

The RIBA Plan of Work is probably the best-known of the process structure guides on design in the construction sector.

First published in 1963, the current edition has been upgraded to include variations on procurement routes and the activities of other parties in the multi-disciplinary team *(RIBA Plan of Work – Multi-Disciplinary Services,* RIBA Enterprises, 2008*)*. In addition there is now a Green overlay dealing with sustainability, and also a forthcoming BIM overlay.

It is critical to have an understanding of the various design stages and what should have been achieved by the design team at the completion of each stage. So often is design information presented as 'Stage E' or 'Stage F', 'E+', etc. to contractor teams. While this may largely be true, deeper examination of the information reveals that critical activities have not been fully completed or sometimes not carried out at all.

Understanding the RIBA Plan of Work as a status of design progress is only a guide to where you are in the process. However, you need to understand what has been left out or is incomplete. These items will become risks as you go forward, whichever side of the table you are sitting on, and the project team's degree of success in resolving the issues will have a great bearing on the outcome of the scheme.

Very often in developing the design through the stages, the Design Team will be aware of areas that still need work or resolution. This may not be immediately apparent, but clearly these issues will 'come out of the woodwork' later and will need to be resolved. In my experience they have ranged from practical constructability of cladding detailing, and compatibility of tolerances and deflections, to ongoing fire strategy issues and so on. The skill lies in identifying these issues early enough to head off any problems, and that is where a DMer can add value through risk resolution.

Therefore it is important to understand the structure of the stages, activities and key outputs. This then enables you to make a quick assessment of design status.

The stages are as follows:

PREPARATION

A Appraisal

- Identification of client's needs and objectives, business case and possible constraints on development
- Preparation of feasibility studies and assessment of options to enable the client to decide whether to proceed.

B Design Brief

- Development of initial statement of requirements into the Design Brief by or on behalf of the client, confirming key requirements and constraints
- Identification of procurement method, procedures, organisational structure and range of consultants and others to be engaged for the project.

DESIGN

C Concept

- Implementation of Design Brief and preparation of additional data
- Preparation of Concept Design, including outline proposals for structural and building services systems, outline specifications and preliminary cost plan
- Review of procurement route.

D Design Development

- Development of concept design to include structural and building services systems, updated outline specifications and cost plan
- Completion of Project Brief
- *Application for detailed planning permission.*

E Technical Design

- Preparation of technical design(s) and specifications, sufficient to co-ordinate components and elements of the project and *information for statutory standards and construction safety.*

PRE-CONSTRUCTION

F Production Information

- **F1** Preparation of detailed information for construction

- *Application for statutory approvals*

- *F2 Preparation of further information for construction required under the building contract*

- *Review of information provided by specialists.*

G Tender Documentation

Preparation and/or collation of tender documentation in sufficient detail to enable a tender or tenders to be obtained for the project.

H Tender Action

- *Identification and evaluation of potential contractors and/or specialists for the project*

- *Obtaining and appraising tenders; submission of recommendations to the client.*

CONSTRUCTION

J Mobilisation

- Letting the building contract, appointing the contractor

- Issuing of information to the contractor

- Arranging site handover to the contractor.

K Construction to Practical Completion

- Administration of the building contract to Practical Completion

- Provision to the contractor of further information as and when reasonably required

- Review of information provided by contractors and specialists.

USE

L Post Practical Completion

- **L1** Administration of the building contract after Practical Completion and making final inspections

- **L2** Assisting building user during initial occupation period

- **L3** Review of project performance in use.

Note – Items in italics can be moved to different stages to suit project requirements.

(*RIBA Plan of Work – Multi-Disciplinary Services,* RIBA Enterprises, 2008).

Salford Process Protocol

The SPP presents a slightly different take on life. It is a whole-life project process, uses less 'designery' language and is an integrated team model. It is possible to take each stage and break it down level by level into sub-processes and routines.

A key feature of the process is gateway reviews. This means that at the end of each stage, all the work activities are reviewed for completeness and to identify any outstanding items or issues.

Presentation of the process, stages, activities and participants is in a more visual form, which is more readily legible (www.processprotocol.com).

Office of Government Commerce – The OGC Gateway™ Process

The OGC Gateway™ process provides a framework for delivery of government-sector projects, and is a mandatory requirement. Similar to the Salford Protocol, the OGC process is more a project-/business-centric process rather than a design-centric process like the RIBA Plan of Work.

The OGC is supported by an excellent website and publications for download: www.ogc.gov.uk

The stages are:

Gate 0 – Strategic assessment
Establish business need. Once a business need is identified, the SRO (Senior Responsible Officer) should consider whether a construction project would be the best way of meeting it.

Gate 1 – Business justification
Develop business case. As the business case is developed, the SRO should ensure that the scope of the project is clearly defined and that opportunities for adding value have been explored thoroughly.

Gate 2 – Procurement Strategy
Develop procurement strategy. The SRO needs to ensure that the most appropriate procurement route has been taken to procure the integrated project team and that the commercial arrangements will deliver value for money.

Gate 3 – Investment decision
Competitive procurement. At this stage the whole-life costs should be clear enough to enable the SRO to commit to a maximum budget; the scope of the project is finalised.

Decision Point 1
Outline design. At this point the SRO/design champion signs off the outline design. After this point no client changes should be made.

Decision Point 2
Detailed design. At this point the SRO/design champion signs off the detailed design.

Gate 4 – Readiness for service
Construction completed. At this stage the SRO confirms that the facility has been commissioned, is fit for purpose and ready for use.

Gate 5 – Benefits evaluation (repeated as required)
Operate facility. The business benefits that have resulted from the investment in the facility are checked against what was expected in the business case. There may be several Gate 5s throughout the life of the facility.

Publications are accessible from the OGC website and there are a number of publications in The Achieving Excellence Procurement Guides, which cover subjects such as Project Organisation and Responsibilities, Procurement Lifecycle, Risk and Value Management, Integrated Teams, Teamworking and Partnering, Procurement and Contract Strategies, Whole-life Costing and Management, Improving Performance, Project Evaluation and Benchmarking, Design Quality, Health and Safety, and Sustainability.

Avanti

Avanti is a collaborative approach for an ICT project environment. Avanti achieves its core objective in three ways: through providing consultants, each an expert in the Avanti approach; information standards and procedures; and cross-project assessment and measurement.

Evaluation of the impacts of Avanti has shown:

- Early commitment offering up to 80% saving on implementation cost on medium-sized projects

- 50–85% saving on effort spent receiving information and formatting for reuse

- 60–80% saving on effort spent finding information and documents

- 75–80% saving in effort to achieve design co-ordination

- 50% saving on time spent to assess tenders and award sub-contracts

- 50% saving on effort in subcontractor design approval.

Source – from the CPIC website: www.cpic.org.uk/en/publications/avanti/

DM and CDM – Health and Safety

It might seem odd to add a section here on Health and Safety, but it is a critical strand of the design, construct and operate processes.

We take our personal safety for granted. None of us goes to work thinking about whether we will return home alive or uninjured.

Sadly, in the construction industry that is still not the case. Every day someone is injured on site, and every few days someone dies on a building site in the UK.

Clearly this is not acceptable; we all need to play our part and this subject is not to be taken lightly.

I suspect that in the past designers have paid lip service to some aspects of the complex projects they design, particularly in terms of construction sequence and maintenance, component replacement and cleaning.

From a DM perspective, in reviewing design information at any stage you need to think of a few key points. Remember to refer to the CDM 2007 ACOP for detailed guidance; remember that it covers construction *and* buildings in use. Safety needs to be considered as much in the design process as you would consider aesthetics, structure, function, space and all the other design requirements. There isn't a coat of paint called 'safety' to be applied to a design afterwards!

The HSE carried out a survey in 2003 on 123 major projects. Feedback indicated that a third of designers had little or no understanding of CDM responsibilities. Only 8 per cent claimed to have received training, and frequently designers were leaving it to the principal contractor to resolve safety issues. Contractors were struggling to control risks on site that could have been eliminated or reduced.

The danger, of course, is that in responding to address these issues, the team produces reams of generic risk assessments that are of no use to anyone. Specific assessments are required that specifically relate to the project and its own particular challenges.

The CDM 2007 ACOP is not prescriptive – it does not tell you how to do it, but does tell what you must achieve. The ACOP tries to move us away from just box ticking and generic assessments to something more specific and useful.

What you must do:

- Make clients aware of their responsibilities and duties
- Avoid foreseeable risks, for construction and future use
- Eliminate hazards and reduce risks
- Inform and coordinate with the project team and stakeholders.

You must be able to show that you (the team) have considered:

- The people who build
- The people who will clean and maintain
- The people who will use the building
- The people affected, customers, public, etc.

Provide information – this all needs to be communicated by:

- Notes on drawings
- Written information
- Suggested construction sequences
- Pre-construction information (Refer to the ACOP Appendix 2).

What designers don't need to do:

- Consider unforeseeable hazards and risks
- Consider possible future uses that can't be anticipated
- Specify construction methods
- Exercise H&S function over contractors.

Things to think about:

- Has a CDM-C been appointed?
- Has the pre-construction (pre-tender) information been provided?
- Have the designers provided design risk assessments?
- As the design develops, are the DRAs updated?
- Is the CDM-C involved in ongoing discussions?

- Are the designers considering construction sequence and logistics in their design approach?

- Are designers flagging issues up by noting them on their drawings?

- Are decisions being recorded somehow, by means of decision trees, audit trails, file notes, meeting minutes?

- At Practical Completion, have residual design risk assessments been provided?

- Has a cleaning and maintenance plan been provided?

Usually designers refer most of this to the contractor. For the normal everyday items that is acceptable, but the designers have a duty to flag up issues that are not run-of-the-mill items. For example, there might be part of the structure that needs constructing out of normal sequence and special propping over and above what is normally required, perhaps to maintain structural stability in a temporary condition. As an example, on an office project I worked on several years ago, the structural engineers had produced a 3D study of the installation requirements for some special precast concrete planks they had designed, requiring a particular installation sequence and additional temporary supports.

I think there has to be a degree of balance and common sense. Designers will always want to push the boundaries in terms of technology and what can be built; innovation in design will always be so, and so it should be. The construction industry responds by finding safe ways of achieving designers' aspirations – so there is creative tension in this. Most constructors I know just love to rise to the challenges!

However, the industry needs to step up in terms of its image and safety is a key part of the picture. It is a basic human right to be able to go to work, work in safety and return home in the evening without mishap.

There are three publications that will be of help:

- CDM 2007 – Approved Code of Practice (ACOP) HSE

- CDM 2007 – Construction work sector guidance for designers, 3rd Edition, CIRIA

- CDM 2007 – Workplace 'in use' guidance for designers, CIRIA.

The work-sector guidance publication is useful as it provides information related to most subcontract packages. Information provided includes scope, exclusions, major hazards, specific hazards and prompts, examples of risk mitigation and also relates the advice to design stages. This can be used as an aide-memoire for reviewing packages, and provides a good start to help you produce a project-specific safety review.

In recent years the HSE has moved away from numeric risk assessments methods to a more descriptive approach. This should itemise the hazardous activity, information provided for hazard control, and details of any residual hazards after mitigation. There is guidance available on the HSE website: www.hse.gov.uk/construction/cdm/faq/designers.htm

Remember that the safety strategy on a project should be developed by the whole team, but the project procurement process may well cut straight across this, as dialogue between constructors and designers may be restricted until after contract award. Under the ACOP, designers have a duty to make relevant information available at tender stage. Constructors will want to understand as much as possible about the scheme during the tender stage to inform their proposals and tender offer.

The sooner everyone can get round the table and discuss the project safety strategy, the better. This is not just about economic or financial loss, but building sites can be dangerous places, and people lose their lives. In the worst possible scenario, if there is a loss of life on your site, as a designer, constructor or Design Manager you could be called to account for your actions by a visit from the HSE and could potentially face criminal charges.

Looking at this positively, regular review of safety in design and construction must be a key part of your DM process, and should form a standing item to be reviewed in every Design Team meeting, subcontractor review meeting, etc. The CDM-C will be delighted to assist with this, to ensure the project team are meeting their CDM 2007 obligations.

DM and Cost Management

Successful projects are the result of a successful fusion of design, commercial and construction disciplines. It is essential that the Design Manager is adequately supported by a quantity surveyor or estimator. However, when the design development process is happening, how do you keep a grip on the cost? Finishing the design information on time may well be an objective, but if when the analysis is completed the cost is found to be way over budget, there will be fallout! In this section we look with Alec Newing at some key principles about design and cost.

The Design Manager needs to work hand in glove with a few key people:

- The designers (obviously)

- The project, package, or site managers who are going to build the project

- The supply chain

- The surveyor and/or estimator, who is building up the cost plan or contract sum, or managing out the contract, depending on what stage you're at.

This is a balancing act. Design, operations and commercial all need to be kept in the loop, all the time, with attendance and involvement at workshops, circulation of meeting notes and minutes, comments, RFIs, approval processes, etc. All of these activities and communications need to take into account keeping all parties informed and up to date.

At the outset of the design process there could be a Cost Plan (CP) in existence; at this stage this might be more or less an outline, but it will nevertheless define the customer's budgetary requirements. It is essential that the cost boundaries for the project are defined, and are developed in more detail as the design progresses. The CP will probably have been prepared by the customer's quantity surveyor (QS); but whether the CP actually reflects the design that it is supposed to be based upon is a key question, and that is frequently where problems can start.

Quite often, for a variety of reasons, the CP will not have kept up with design development, or it may not reflect all of the design intent. Divergence between the CP and the design information upon which it is supposed to be based can potentially lead to problems further down the project process, possibly leading to emergency Value Engineering (VE) or project and budget changes, which can be embarrassing for the team and costly for the customer. Ideally VE should be avoided unless significant circumstances are dictating change. Too much VE results from the combination of poor DM and cost management. A much better mindset is for value management to be a continuous strand of the process as the project progresses, which is discussed in more detail in Section 6.

Cost-plan updates must be completed regularly and at least prior to the end of each design stage. It can be argued that a design stage is not complete without a cost-plan check, and that the design solution and cost must align at the end of each stage – otherwise you cannot effectively evaluate contingency and design development. The end of a stage presents the opportunity to rein everything back in before shooting off on the adventure of the next stage. Otherwise, the design/cost conversation can quickly spiral out of control.

At the cost-estimating stage, what is not on the drawings is potentially more important than what is actually shown. The real skill of an experienced estimator or surveyor lies in seeing beyond what is shown on the drawings and reading in the implications, making allowances and costing accordingly.

The DM can assist the estimator in assessing the content of the design still to be completed, and something that is frequently missing is the design development funds for this – *a cost plan for the design that is still to be completed*.

How is cost-plan data best conveyed to designers so that they can understand it? Rather than simply relying on a spreadsheet, it can be conveyed by citing examples, pictures, sketches and commentary. Perhaps it is more that the CP is a team conversation, as opposed to just a document – i.e. it needs interpretation and explanation. The conversation should happen before the design solution, not after the design has moved on.

In managing cost and design there are a few key tools and techniques that can be used:

Benchmarking data

Most cost consultants and contractors will have a good idea of the cost rates for particular types of construction, elements and buildings. This will usually be in

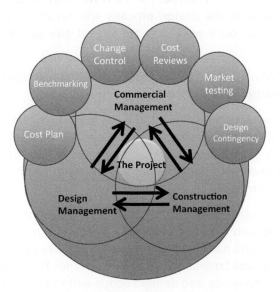

Figure 2.6 Commercial Management integrated with DM and Construction Management – any successful project is the result of effective collaboration and integration between all disciplines.

the form of cost per square metre or cubic metre, or by quantities. Coupled with this rate there will be various assumptions attached about quality of specification and types of component, as well as what is excluded, such as abnormal ground conditions etc.

Therefore it should be possible at an early stage to produce rough costings for the project, but these must be linked with the key assumptions that the costing relies upon. An estimator will then build up the Cost Plan to include allowances for abnormal factors such as ground conditions, and also site set-up, overheads and profit, etc.

It is critical that the design information that the CP is based upon, at that stage, is collected together and incorporated with it. The CP, at whatever stage it is, is only ever a snapshot taken at a particular moment in the project's life. As the project design progresses, further snapshots will be taken. It is crucial that you can articulate the differences between the design/CP at the different stages.

This leads us on to consider change control.

Change control

Design changes and develops as the project progresses. Designers are trained to iterate, refine, develop, enhance, improve and continue the design process. Clients change their minds. Circumstances and situations change. Economic factors move with the markets and financial institutions. All of these and many more factors can impinge on the design process, resulting in change. As soon as something changes, then the cost plan is out of date, and potentially successful delivery of the project is in jeopardy.

Therefore having a change-control process in place is critical to your delivering to cost. There is a reluctance to grasp this sometimes. It can seem complicated and so teams avoid it, whether it is because of the discipline required to impose it or the level of scrutiny required to make it work. However, a changing design, with an out-of-step cost plan progressing unchecked, will lead to difficulties later on. It is better to square up to this at the start and put procedures in place to avoid difficulties later.

As the RIBA Plan of Work points out in various places, changes to the design after certain points in the design process will result in abortive work and costs for the team, and once work has started on site possibly for the contractor (and customer!).

It is important that the whole team, including the supply chain, understands how change control is to be managed on the project.

Changes initiated by the client or other team members must go through a sign-off process, whereby the change is identified, scoped and evaluated for any implications such as additional costs, and effects on the programme. If the change is approved and implemented, then that needs to be officially instructed to the team by the customer.

An area of ambiguity requiring careful management is the process of commenting on subcontractors' shop drawings by the design team. At this point in the project, the contract will have been let; you will probably be on site and the contract sum, programme and targets will have been set for the project. Each subcontract will have a cost envelope and defined scope within which it needs to deliver if it is not to go over budget. Cost, quality and time have therefore been defined in some form and detail at this point.

Frequently in the process of reviewing drawings, designers will add comments that perhaps add to the scope, possibly enhance the quality of specification, or change the detailing. Potentially all of these can add to the cost and have their implications that affect manufacture and ability to deliver within the agreed times-cales. This process of 'design creep' often occurs and costs can escalate quickly if unchecked.

To remedy this, make sure that meetings with subcontractors include review of the design and the corresponding commercial status. Make it clear to designers that they must keep within the defined scope and cost parameters. If change is required, then it must go through the change-control process. Remember that this process of design development through subcontractors can quickly get out of control – so all the drawings and information shuttling between designers and subcontractors must be thoroughly reviewed, together with any added comments, to ensure that the package design remains within the project cost, time and quality boundaries.

Reviews

In the period between developing the first cost plan and full market testing (see below), then the design should be reviewed as it is being developed by the design team.

As noted above, as the design is developed and evolves, it is important to keep in step with the cost planning. Involve your commercial team as much as possible to make sure that they are aware of any changes in design, detailing, scope and specification every step of the way. Changes to the design and the updated cost plan need to be reported regularly.

Ideally, all workshops and design meetings should have commercial involvement. This shortens the lines of communication and avoids inaccurate transmission of information. Likewise it may be necessary to include the Project Planner and site managers to review programme and construction implications.

Market testing

This is the process of sending out subcontract tender packages or sections of work. It could be an informal process, whereby a few subcontractors are brought in to meet the Design Team to discuss ideas and costings, commenting upon the design information and offering advice; or the more formal process where documents are sent out to a list of tenderers with formal tender returns.

The subcontract tender documents will consist of a package scope, specifications, drawings, possibly including Bills of Quantities and also any business-specific subcontract forms.

Tender returns from subcontractors in response to package enquiries must be reviewed carefully. Rarely will they be totally compliant with the package require-ment documents. Hidden in the small print could be substitutions on quality and performance, or limitations on other aspects, which could affect the cost of the particular package. Implications might be that the preferred subcontractor may not be capable of being compliant, in which case the variance will form part of the Contractor's Proposals.

Also check that interface issues with other packages have been included and costed, in accordance with the project requirements.

The consequences of not reviewing properly the returns and information at this stage will result in sudden cost additions or changes later, probably on site.

Elements or items perhaps not costed or included properly at this stage will become apparent later, making life difficult and potentially embarrassing for everyone (you can imagine the response you might get: *What? Didn't you read this? Didn't you understand the implications?*) – leading to last-minute changes, or additional work and cost to enable the project to proceed.

In my own experience on a project, a cladding subcontractor downgraded a glazing specification in their tender return. This did not come to light until we had to prove compliance with the City of London bomb-blast requirements. We resolved the issue, but there was an additional cost – embarrassing!

The estimator or surveyor will take the package tender return and reconcile it into the overall cost build-up, adding on attendances, allowances, overheads and profit, etc.

The DM's role in assisting with this process is crucial if the costing is going to truly reflect the project as designed, so DMer comments on compliance with the requirements, interfaces, missings or substitutions in package returns are important. In some cases this information forms part of the Contractor's Proposals, which we will look at in the Procurement section.

Design contingency

This is the risk budget for 'unknown unknowns'. It should diminish as the design develops and will be closed out only once the project is built and has run through at least a couple of operational cycles, as there could always be additional design work required to get the facility working as it should.

This is different from design development monies, which are for 'known unknowns' – such as items that we know will need to be designed – steelwork connections, for example, that have not been closed out yet. Design is still to be completed, as I discussed earlier.

When does design development become design contingency? When the design to be developed cannot be resolved within the Design Development budget set – something unforeseen has occurred, therefore the Design Contingency should be called upon.

Who owns these budgets and has accountability for them is the key issue. It should be the DMer, who is needed to ensure that the design can be managed in an agile way. The commercial team should have the responsibility for preparing the budget and cost information, as that is within their sphere of competence – 'able to respond = response ability'.

For some, management of risk is simply allocating allowances against activities or operations. In theory that might tick the box, but in a competitive market, an over-provision of risk monies could be the difference between winning and losing a tender. It's much better to have a realistic view of the team's abilities to manage risks, which needn't necessarily mean additional cost – it's more that people just need to carry out their roles effectively.

So in summary:

- Use benchmarking data – be aware of assumptions and exclusions.

- Use a change-control process to ensure the Cost Plan keeps pace with design development.

- Hold regular formal reviews, include commercial, planning and construction representatives.

- Be fully involved in market testing with the commercial team, review returns, look for the missings, gaps between packages, and substitutions on specification or performance.

- Beware of 'design creep' on packages, enhancement and scope.

Last Planner

(by Alan Mossman)

Last Planner: more predictable, more reliable projects

You may have heard of the Last Planner, which applies 'lean construction' principles to the design and construction processes – maximising value and minimising waste. This is perhaps as much a tool as a process, but I've included it here as it would connect with your planning the DM process for your project. There is considerable synergy in the connection between lean thinking and BIM, which will lead to lean thinking becoming much more easily integrated into project processors.

The Last Planner® System (LPS) is a commitment management system used to create predictable and reliable production in design, construction and other one-off, project-based production settings. Developed by staff of the Lean Construction Institute in the USA (www.leanconstruction.org), it builds trust and uses peer pressure as a critical motivator.

The Last Planner System ensures that project staff have the right conversations at the right level at the right time to enable the smooth delivery of their project. There are five critical conversations – in black in Figure 2.7.

The milestone schedule is the work of the first planners – the professional planners, using project management software such as MS Project and Primavera – and is part of testing the feasibility of the client's delivery criteria.

It is the *last planners* (trade foremen on site, design-team leaders) and project managers who:

- produce the overall project schedule collaboratively so that they understand the total process before they start

- work together week by week to ensure that work **can** be done when required and to plan production

- continually improve both planning and production.

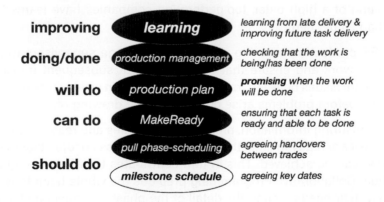

Figure 2.7 The Five Crucial Conversations. Reproduced by permission of Alan Mossman, The Change Business Ltd.

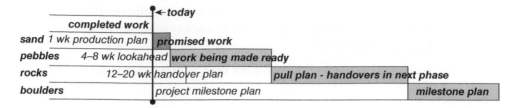

Figure 2.8 As the date for work to be done nears, the level of planning detail increases. Reproduced by permission of Alan Mossman, The Change Business Ltd.

The last planners are generally able to produce a schedule that is significantly tighter than that produced by the first planners – 20 per cent faster is not uncommon. Pull-based phase scheduling and the systematic MakeReady process enable last planners to make commitments or promises for the work they will do the next week (or day) in the context of the production planning meeting. Their success in delivering those commitments – measured to the day, and more recently even to the hour – is tracked with a commitment reliability metric for the project as a whole (often referred to as PPC – percentage of promises completed).

With a small number of easy-to-complete forms, LPS is simple to administer using Post-it® notes, paper, pencil and a photocopier; MS Excel® or a similar spreadsheet can help. On larger projects there is web-based software available to support the MakeReady production planning and learning conversations. Software can never replace the conversations 'at the wall' as the last planners agree the programme that they will build to.

Figure 2.8 shows how work packages are progressively broken down as the date for production nears.

Last Planner is based on a strong theoretical foundation and has been subject to considerable research scrutiny. For many organisations it is a first step on the journey to lean, as it helps to stabilise production processes, but you don't have to be lean to get some valuable benefits from it.

Benefits

When teams use LPS the reliability of production plans improves. Profit and productivity grow with plan predictability (measured in terms of 'PPC' – the percentage of work promised for a given day delivered on time and to quality). Where LPS is not used, PPC measured to the day is generally below 40 per cent. Scores of over 70 per cent show projects making good money. With integration and collaboration of a high order, top-performing companies have learned to achieve more than 95 per cent.

LPS production-planning processes encourage supervisors to plan and prepare their work and ensure that they know what subsequent trades expect of them. Critical issues between trades are worked out in weekly production-planning meetings before problems arise on site or in the drawing office.

By ensuring information, prerequisite activities and resources required to perform a task are ready in time, the MakeReady process ensures that the planned production can be achieved. This in turn contributes to safer working and waste reduction. Collaborative Programming prepares the whole team to work together; they get their heads around the detail of the phase or project and agree how they will assemble the building or create the design; anticipate problems and identify unresolved details; see opportunities to reduce cost and or improve the quality; consider risks and develop countermeasures where they can. Programme Compression can reduce the length of a programme – one contractor took 6 weeks from a 19-week programme.

Generally LPS is linked to improved safety (65 per cent fewer accidents in one study in Denmark, 75 per cent in another in Chile), reduced sickness absence, improved quality and delivery and a smoother-running site with less firefighting.

We do it already.

Most project managers do some or all of these things some of the time.

Last Planner is a formal and disciplined execution of a system of related elements. Major benefits come when the whole project team uses all the elements rigorously over time. The greatest benefit comes when an integrated team uses the elements consistently over a number of projects.

Signals that you are not *yet* doing it include:

- work being pushed into production by the programme
- continual firefighting
- subcontractors with no sense of ownership of the programme
- work being done out of sequence
- operatives with little or no sense of what they will be doing the day after tomorrow.

When he was with BAA, Gerry Chick spoke of the value of getting everyone round a table and he welcomed the recognition LPS gives to bad news. 'Bad news,' he said, 'provides good information. Bad news *early* is even better.' Last Planner enables bad news to surface before it becomes a major issue and can also provide signals of imminent crisis that may enable the team to head it off.

BAA is one of a number of organisations that require the use of Last Planner on their contracts.

To sum up – A few thoughts

Business process

Whatever company you are currently working for, there will be company processes that relate to project delivery and design management that you, as the DMer, have to comply with.

So while the above processes will inform what you should do, you need to ensure that whatever outputs you produce comply with your business's requirements.

Processes in conflict

I am now going to make a couple of wild generalisations just to make a point – but I am sure you will understand where I am coming from.

Design process is essentially iterative (repetitive), going round and round, diving off into dead ends, returning only to dive off again. Hopefully the iterations are moving forward and coming together, but sometimes as options are explored the process can move backwards for a while. There are lots of creativity, intuition and ideas, and sometimes these are not necessarily worked out in full detail at any given point in time.

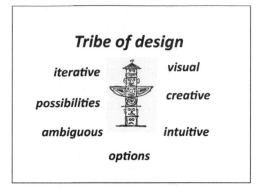

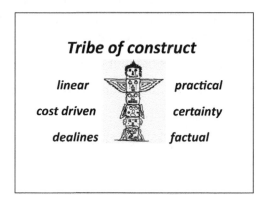

Figure 2.9 Two Tribes.

Construction process, in contrast, is essentially logical, linear, with acute attention to detail. After all, the building has to be built. Doing this are really practical people who just want to know what to do, what to build and to be able to get on with it, as quickly and effectively as possible. Time is money!

Two tribes

There is the tribe of design, and the tribe of construct. They both want to do a great job. They both want to deliver successfully for their customer.

But they work differently, and think differently. The two processes are fundamentally in conflict!

As a DMer you need to remember this fact, as it is your raison d'être. As one of my previous bosses put it: 'Designers speak Greek, contractors speak Latin, Design Managers speak both!'

The role of Design Management is to act as the interface between design process and the construction machine. The DMer ensures that the outputs from the iterative design process are delivered in the right form, at the right time, to enable the construction machine to continue its course down the project highway, without deviation, interruption or any other issues getting in the way.

Information flow

BIM will change many things, but one good thing is that, in the right hands, it should drive efficiency of information flow. Currently, design teams push out information by the bucket-load: hundreds and thousands of drawings and revisions over the lifetime of a building project. Down the information chain, it goes on even more. How many times has a surveyor asked me to send a subcontractor everything we have in order to close out the risk that they might not receive the right information if we're more selective? This is not only information overload, but also information and resource waste. Multiply it by all the consultants, subcontractors and contracts in the UK alone and it translates into millions of pounds, and days of wasted effort and information.

The drawings are the outputs, and sometimes people are producing drawing after drawing without really thinking about what they are doing. Who is this for? What do they need to know? Why do they need to know it? What will they do with it? How will they use this information?

CDP subcontract packages are a good example. Quite often professional design teams virtually design out the package before handing it over to the subcontractor – only for the subcontractor to redesign it, because their system isn't like that, they don't use that kind of fixing, or their specification is slightly different and so on.

At the heart of this really is communication about what members of the team do, and what they need in order to do it. I suspect that if designers understood more about what subcontractors do, they would produce design-intent drawings that contain less unnecessary information, but would be more relevant and targeted to their purpose.

This is only one example, but generally we all experience information overload on a daily basis. We need to consider why we are producing what we do.

In a BIM project environment the data model is the deliverable. Members of the project team access the model to obtain whatever information they need. Note that this moves us from a **PUSH** to a **PULL** dynamic. If you need information, you just go and get it for yourself. There will be no need to rely on an information manager distributing information to the project team. BIM changes a lot of things, including the dynamics of project process. Information flows have the potential to become more pertinent, accurate and focused.

And finally for this section . . .

We have considered a number of process models that as DM you will come across and you will need to be aware of.

- *Interested in understanding design status and stages? – RIBA Plan of Work*

- *Integrated team scopes and services? – CIC Scope of Service*

- *Public-sector procurement? – OGC Gateway™ Process*

- *Information management in a BIM environment? – BS1192:2007*

And so on.

In considering your own situation you will draw upon the industry processes and models we've discussed (and others), taking what you need to synthesise them with your own business/organisation requirements to meet the objectives of your project.

So having considered some processes, we now move on to looking at some tools the DMer can use to keep control of the design process.

---

3 DMTCQ

The CIOB Design Manager's Handbook

The CIOB Design Manager's Handbook

Connect – Integrate – Communicate – Innovate – Collaborate

DMTCQ: The CIOB Design Management Benchmark

A generic framework to manage design

As I stated earlier, Process and Tools are the 'engine room' of Design Management. I have placed the DMTCQ here in the Handbook as it sits naturally between these two sections.

Managing design successfully can be accomplished with a core set of tools, and this process of management has certain basic characteristics, or 'hallmarks', irrespective of the project type, size, sector, business, organisation, etc.

Just think about it for a moment – we come together as a team to design and construct something for our customer. There are a number of parties involved; information needs to be produced and exchanged in various forms; suppliers, manufacturers, constructors and other specialists need to be engaged in the process – all to deliver the end product, at the right time, for the right cost and providing and meeting the right quality.

How can you tell that Design Management is taking place and is working? How do you know? What is the proof? DMTCQ defines activities and processes that should be in place, and can be checked.

These principles apply equally to a road, a ship, a nuclear power station, a school or someone's house.

> If design is happening, and is being managed effectively, then certain key things need to be happening as a consequence and the results of those management activities should be both visible and tangible in some shape or form.
>
> There should be visible outputs or outcomes – 'the Hallmarks of DM'.

In this section I have brought these 12 key hallmarks together, which I have called 'The DM Benchmark' or 'DMTCQ' for short (based on Time–Cost–Quality). This brings together several tools and processes, which should enable you to control

The Design Manager's Handbook, First Edition. John Eynon.
© 2013 The Chartered Institute of Building. Published 2013 by Blackwell Publishing Ltd.

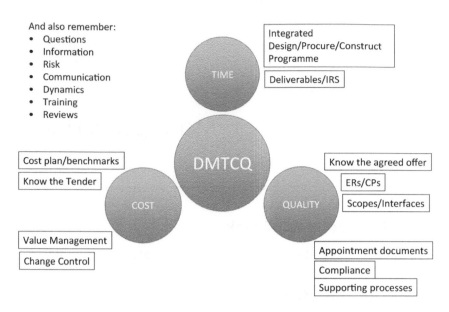

And also remember:
- Questions
- Information
- Risk
- Communication
- Dynamics
- Training
- Reviews

Integrated Design/Procure/Construct Programme

Deliverables/IRS

DMTCQ

TIME

COST

QUALITY

Cost plan/benchmarks

Know the Tender

Value Management

Change Control

Know the agreed offer

ERs/CPs

Scopes/Interfaces

Appointment documents

Compliance

Supporting processes

Figure 3.1 DMTCQ – A framework for Design Management

the design process. Conversely where a design management process is not delivering the results as expected, then you should be able to diagnose what is not working and why, so as to get the process back on track.

Remember there is no magic or 'silver bullet' for DM. Methodical and consistent application of strategy, management, delivery and definition of objectives and outcomes will pay dividends. Good DM is simply good management practice and the use of appropriate techniques.

Note that these are only principles. I have not been prescriptive about the exact 'How', as this will vary from project to project, depending on sector, size, complexity and timescale. For example a programme of activities and milestones could be produced as a bar chart, or a schedule of dates. What is important is that you, as the DMer, are managing the particular aspect, using an appropriate tool, and you are therefore able to demonstrate that you are managing that aspect effectively.

DMTCQ: The CIOB DM Benchmark

This is a suite of basic tools and processes to control DM on a project and to enable you to understand how you are doing, based on Time, Cost and Quality: TCQ, 'The project envelope'. Refer to the Four Stage Process (see Figure 1.1).

All the tools and processes shown below need to be in place and managed effectively to provide 'the Hallmarks of DM' that I have discussed earlier.

Time

- **Programme** – Design/Procure/Construct Programme – fully integrated, and agreed with the whole team. Monitored, reviewed and managed

- **Deliverables schedules/IRS** – in place and agreed with the whole team. Monitored, reviewed and managed.

Cost

- **Cost plan/benchmarks** up to date with the design iterations, reporting in place

- **Tender** – know the full details of the agreed offer, including any clarifications/exclusions – understand any particular conditions

- **Value management** – are there any potential ways of improving the value proposition of the scheme for the customer? Consider design, delivery, life cycle

- **Change-control** management procedure – TCQ implications of any design change, and to be formally reviewed and instructed.

Quality

- **Agreed offer** – know the details of the agreed offer – understand what has been agreed to in terms of function, content, standard, cost and timescale

- **ERs/CPs** – Employer's Requirements/Contractor's Proposals – understand the relationship between these documents, and the requirements

- **Scopes/Interfaces** – understand the scope, content and interfaces of the Works Packages, check for gaps, overlaps, duplications, inconsistencies

- **Appointment documents** – understand exactly what the consultants are contracted to do; is everything covered? Understand their resources and capabilities. Are their services and resources sufficient and appropriate?

- **Compliance** – planning, listed/conservation consents, Building Regulations, other code requirements, processes in place to achieve approvals

- **Supporting processes** – in place for CDM/Safety/Sustainability/Environment/Project-specific.

And remember:

- **Questions** – always have clarity of role, objectives, scope, accountabilities, for all parties. Think of the Who, What, When, How, Why?

- **Information** – always know all the design information – keep the drawings, specifications, schedules, information, under constant review

- **Risk schedule** – be aware of the risk-management processes, and ensure that risks are eliminated, reduced and managed effectively

- **Meetings/communications/reports** – make sure that communications and reporting are working effectively at all levels

- **Dynamics** – think about people issues. Are there personality issues/agendas driving situations?

- **Training** – do any parties/individuals need training on processes, technology or other project aspects?

- **Stage/gateway/interim reviews** – review the work at key milestones and stages, identify activities and outputs that are not complete or not up to the standard you need, have a strategy in place to deal with outstanding issues before starting the next stage.

The above Processes and Tools are expanded in more detail in Section 2: Process and Section 4: Tools.

4 Tools

The CIOB Design Manager's Handbook

I keep six honest serving-men
 (They taught me all I knew);
Their names are What and Why and When
And How and Where and Who.
I send them over land and sea,
 I send them east and west;
But after they have worked for me,
 I give them all a rest.

I let them rest from nine till five,
 For I am busy then,
As well as breakfast, lunch, and tea,
For they are hungry men.
But different folk have different views.
 I know a person small –
She keeps ten million serving-men,
 Who get no rest at all!

She sends 'em abroad on her own affairs,
 From the second she opens her eyes –
One million Hows, two million Wheres,
 And seven million Whys!

– Rudyard Kipling
The Elephant's Child

Introduction

In most companies there will be a QA (Quality Assurance) system, possibly registered to ISO 9001 and/or 14001 (Environmental Management Systems).

This will mean that there are procedures to follow, probably forms and checklists to use, which will be regularly audited externally to ensure that your company retains its registration.

It's not to say that forms and checklists aren't useful, but like most things in life it's a question of balance.

Tools, processes, IT and software are all here to help us get the job done; they are not ends in themselves – although the people in charge of tools and processes (and IT) sometimes forget that!

The Design Manager's Handbook, First Edition. John Eynon.
© 2013 The Chartered Institute of Building. Published 2013 by Blackwell Publishing Ltd.

The tools we use need to be appropriate to the task and readily usable. If they are too complex, too large or bulky, and intimidating, people won't use them and the consequent result is effectively anarchy and inconsistency of delivery – coupled with the very real possibility of catastrophic project failure.

So producing multi-page spreadsheets and manuals that rival 'War and Peace', while perhaps containing the right information that will rarely be used properly because they're just too unwieldy, is in my view just a waste of effort. There has to be a balance in applying the right techniques and keeping it light enough so that people will actually use the tools in order to reap the benefits.

This is why I have not adopted a tick box/checklist approach to DM.

In this section I have listed some tools. Most will be familiar to the professional DMer. I have also included some subjects that perhaps are not exactly viewed as tools, but of necessity they are factors that the DMer has to be aware of in managing the design process. This seemed the best place for their inclusion. Most of them will be applicable to most projects, but not all.

Of course, there are levels of understanding, application and complexity.

At the simplest level, you can be told to 'just do this and this with that and follow these steps', where understanding is not necessarily required. This is the tickbox/ checklist mindset: just follow the instructions. But life rarely goes exactly to plan, particularly where people are involved, so you need to learn to improvise a little between the lines. At this point it helps if you have some understanding of the principles behind what your immediate goal is, and why we use particular tools or techniques in certain situations. If you understand the underlying principles of what is happening, your chances of success will increase when applying processes and tools in managing the design process.

As an industry I think we tend to overcomplicate. Sometimes the simplest, most obvious questions are the most profound. A little naivety occasionally can be a good thing. So keep asking 'Who, What, Why, How, Where, When?'

I have long held the belief that managing design successfully can be accomplished with a core set of tools, and that it has certain basic characteristics or 'hallmarks', irrespective of the project type, size, sector, business, organisation, etc. If design is being managed effectively, then certain key things need to be happening, and the results of those management activities should be both visible and tangible in some shape or form.

In the previous Section 3 I have brought these 12 key hallmarks together, and have called them 'The CIOB DM Benchmark', or 'DMTCQ' for short (based on Time–Cost–Quality). This brings together several tools and processes, which should enable you to gain control of the design process and diagnose where things are going off the rails if it gets to that point.

And so, on to the **Tools**, in no particular order of importance other than alphabetical for ease of reference:

Animations, fly-throughs

Most people will have experienced this aspect of BIM – an animation or visualisation of a project, sometimes as a simple walk-through, sometimes set to the programme timeline (4D BIM) to illustrate the construction sequence and logistics.

Animations can be used as part of a project briefing to, say, new team members or the supply chain, and are a good way of getting the basics about a project across, or explaining to a potential customer what you can do. Many contractors also are using them as a presentation and interview tool to win work.

Designers and some major contractors have in-house teams capable of producing animations, but there are many external companies that specialise in 3D visualisation.

If you are commissioning an external company to produce an animation, then you need to be clear about why you need the visualisation, what is it for, who will see it, and what information you have available. The clearer the briefing you can give, the better the output will be.

The timescale is critical, so be clear about when you want to see a draft version, and when you need the finished product. Check that you have the hardware and the software to play the animation. Technology is always a bugbear! Several times, failure of the technology has resulted in last-minute improvisation! So test, test, and test again to be sure.

Appointment documents

These are the contracts that set out the legal arrangements for each consultant or other specialist to carry out their work on the project.

Most institutes have their own model form appointment documents. These will usually be made up of what I would term a legal 'front end' followed by a number of schedules and appendices. Only the CIC produces an integrated suite of appointment documents that cover all the disciplines.

Where the contractor is going to be employing the designers, either through novation or direct contract, then there are a number of items that you, as the DMer, will need to consider.

Frequently a customer will have had their own appointment documents drawn up by their solicitor, either from scratch or based on a model form. These need to be treated with caution, and will often be very onerous. Check also that if several disciplines have been appointed, their conditions and services provide what is required as a team – there could be gaps in their services, which may require filling via a subcontract package, for example.

Schedules/appendices that should be included as a minimum are:

- Integrated Design/Procure/Construct programme
- Deliverables Schedule and Information Required Schedule
- Schedule of Services
- Matrix of Element or Package Responsibilities.

The intention here is to leave the consultant in no doubt as to what they need to produce, and when. The more detail that can be provided at this stage, the less ambiguity there will be later and this will pre-empt some issues arising.

A few points to check at the appointment stage:

Professional Indemnity Insurance – Check that the level offered/required is appropriate to the size of project. Also make sure that is it offered on the basis of 'each and every claim' and not on an aggregated claims basis, which effectively caps the insurer's liability but gives you reduced cover. Some consultants will attempt to write a liability cap into their contracts, which again needs to be considered carefully depending on the size of project.

Resources – There may be key members of the consultant's team that you feel are essential to the success of the project. It is usual to write into their contract that these team members cannot be changed without the contractor's approval.

Fees – Check that the fees stated cover the scope of services that you require. Fees can be calculated simplistically using a percentage rate and stages: a 'top-down' approach.

The better way, to provide more certainty, is to work out a fee based on the actual deliverables, resources being used and the time required, or 'bottom-up'. This is one reason why the schedules are useful, as you can check whether the fee realistically covers what actually needs to be done. Take this a step further by meeting their teams and putting names against the deliverables. Part of your assessment of the consultants should include whether their team members are sufficiently experienced, with appropriate skills to deliver the design information you need. And naturally the schedules can also be used to measure and manage performance!

You should be able to obtain guidance on fees from the various institutes. There is also a very useful suite of documents published by The Fees Bureau (Mirza and NaceyReseach Ltd) – www.feesbureau.co.uk. They survey fee levels in the UK by discipline, sector and procurement route, and publish these as guides. This provides a useful check on any fee proposals you may be reviewing. Details are obtainable from their website as above.

Novation – That is when the customer has appointed the design team, and then after Contract Award, the contractor effectively takes the place of the customer in the consultant's contracts. A Deed of Novation sets out any changes required to the appointments.

Again, check for all the above headings – the services and details that you need as contractor may well be different from what the customer needed to get out to tender. At this point it is the most straightforward way to resolve any additional services as well as renegotiate any fees.

For example, the structural engineers may not have included reinforcement detailing as part of their scope. Or the designers may have only been appointed to take the project design to Stage E. You will need to review what they have included, and decide whether that is enough for your needs.

Audits

This may mean reviewing the designer, or just as frequently reviewing the design information.

Reviewing the Designer

It may be that you are appointing a design consultant for a design-and-build contract. It might be that you would want to want to get several practices to pitch for the work. Alternatively it may be that you are looking at a consultant who is going to be novated over to you.

So what are you looking for?

Things to think about are:

Experience – How experienced are they in this type of project? Do they have a demonstrable track record?

Resources and people – At this point you should have some idea of the required Design and Procurement Programme, together with the information deliverables. Does the practice have the resources to do what you need? Do the people in the team have the experience to do what is required? What are their qualifications? – and so on. Do you need to supplement their team, with in-house or external support from another practice?

In-house systems – What are their procedures for Quality Management, e.g. is QA registered to ISO 9001? How is work checked before it leaves the office?

Approach to safety – How do they deal with their designers' duties under CDM? How do they communicate this to the team and the principal contractor? What training and safety resources do they have access to?

Approach to buildability – How do they ensure their designs are readily buildable? In their teams, do they have people experienced in construction and site delivery of design?

Experience of working on the procurement route – Whatever form of building contract you are working under, have they done it before?

IT systems compatibility – With BIM this will become increasingly important. What systems do they use, and what protocols? Are there team compatibility/interoperability issues? (Refer to the PIX Protocol (CPIc) at www.cpic.org.uk/en/cpix-on-line-tools/pix-protocol.cfm.)

Professional Indemnity – What is their level and type of cover? Any history of claims made? Details and circumstances?

Financial standing – Dunn and Bradstreet can provide a financial standing report. If unavailable, ask for 3 years' audited accounts.

References – Take up relevant references with clients and contractors.

Note to DMer: Don't forget to review subcontractors with *design* responsibility. Their capabilities need to be reviewed just as much as those of the consultants – you will be relying on their subcontract expertise to turn the design intent into constructed reality.

Reviewing the Design Information:

There are key points at which the design information needs to be formally reviewed.

Ongoing review of all information should happen through regular workshops.

However, at certain points – such as at tender, or when commencing the design development and subcontract procurement process – then you need to know where you are, so you can know how to reach your objective.

As an overview, design information needs to be reviewed for completeness, compliance and cost.

Completeness – Does the information convey all that is needed? Is it of sufficient quality for the purpose, say, if you need to pass it on to a subcontractor or for manufacture or installation?

Compliance – Is the information compliant with the Employer's Requirements and Contractor's Proposals? Have the Subcontract Package scopes been adhered to? Has approval been obtained under statutory regulations, such as Building Regulations? Is planning approval in place? Are there any outstanding conditions?

Cost – Does the information comply with the Cost Plan? Have any changes been made that alter the scope, enhance the quality or add to the cost, and that have not been tracked?

When design information is issued for tender, various statements will be made about the status of design, Stage E, Stage E+, etc., and degree of co-ordination or not. At this point it could be worth investing in a formal audit, as to review a multi-discipline design package fully and in detail will be beyond the competence of most design managers.

A technical audit will cover a number of headings, but again it is completeness, compliance and cost that are the guiding principles.

Key questions include:

- Is the design at the stage it is claimed to be? A review against the RIBA or CIC deliverables helps here.

- Are there any compliance issues such as Building Regulations, fire strategy, BREEAM/CfSH?

- Are here any issues with buildability and construction sequence?

And so on.

This is your starting point for the design development process, so it is critical to understand where you are! – that is, to build up a picture of the current design status, to understand what issues there might be and what actions can be taken.

These comments might be built into a tender clarification schedule or risk register, or might form an ongoing agenda for the design development process.

BREEAM

(Building Research Establishment Environmental Assessment Method)

Refer to the BREEAM website, www.breeam.org, for further information.

Most people by now will be familiar with projects achieving a particular BREEAM rating.

The project is assessed during the design stage and at completion under a number of headings, as below:

- Management
- Health and Well-being
- Energy
- Transport
- Water
- Materials
- Waste
- Land Use and Ecology
- Pollution
- Innovation.

The BREEAM rating benchmarks for new construction projects assessed using the 2011 version of BREEAM are as follows:

BREEAM Rating	% score
OUTSTANDING	≥85
EXCELLENT	≥70
VERY GOOD	≥55
GOOD	≥45
PASS	≥30
UNCLASSIFIED	<30

Refer to www.breeam.org.

A specialist BREEAM assessor will usually be appointed to work with the design team, or sometimes the MEP designer will be employed to do this aspect.

In situations where it is required to improve the rating score as part of the design development process, then the way forward is to identify the make-up of the current scoring, and see also where the additional BREEAM points could be scored if the design were changed.

Usually these potential alterations will incur a cost and/or raise other design or construction issues. A technique to managing this aspect as part of the design process is to develop a schedule (tracker) that can be used to monitor where potential improvements can be made and the required actions and possible costs and implications, including programme, can be managed.

These can be regularly reviewed with the design team and the customer, and if the decision is made to implement the changes, these people can be instructed.

BSRIA Framework for Design Services

Refer to the BSRIA website: www.bsria.co.uk.

BSRIA have produced a guide and matrix for MEP building services design. Included as part of it is a design activities and deliverables matrix for allocating responsibility, which is particularly useful for clarifying design responsibilities between the MEP consultant designer and the MEP package contractor. The latter usually progresses the detailed design and carries out the installation – albeit quite often the consultant is employed by the subcontractor to complete the detailed design.

Building Regulations

Not exactly a Tool or a Process, but nevertheless something that cannot be avoided!

In 1974, when I started training as an architect, the UK Building Regulations were equivalent to the size of an A5 notebook. There are now 19 approved documents, plus references to numerous EN and British Standards and other technical documents, which take up several volumes.

While it will be difficult to know every regulation, you will need to have a working knowledge of the overall requirements. Unlike the planning development control process, where the project is subject to the vagaries of individuals on committees, the Building Regulations represent technical standards that must be complied with. Ensure that advice is sought early in the design process. Achieving

4 Tools

compliance later in the project can result in abortive design work or, once on site, costly changes to built work.

Building Regulations compliance will be dealt with through the Local Authority Building Control department, or by using an approved inspector.

Frequently a Conditional Approval is issued. This will contain a number of conditions to be discharged by the submission of further information. In most cases that information is to be supplied by the design team, and sometimes by particular subcontractors.

This process of discharging conditions must be managed to ensure that conditions are discharged, and the design is approved, before work is ordered or installed. Otherwise the finished work will be at risk, and the Building Inspector may require changes once the work has been installed.

The simplest method of dealing with this is to compose a schedule (tracker) listing:

- The conditions

- Information required to discharge

- Who is the owner/or lead on the condition?

- Who will produce the information?

- When will the information be completed, ready for submission?

The schedule should be reviewed regularly at design team meetings until it is completely discharged.

Carbon emissions and energy modelling

You might want to take a look at this website; www.CarbonAction2050.com. It contains information about the UK carbon-reduction commitments and carbon issues relating to the built environment.

This is an aspect of building design that has become increasingly important. The UK carbon-reduction targets have been set to cut emissions by at least 34 per cent by 2020 and 80 per cent by 2050 – below the 1990 baseline.

Also as energy, oil and resource costs continue to rise, the operational costs of running their buildings will become an even bigger factor in customers' requirements.

An assessment of the design can be carried out, usually by the MEP design engineer, using a software package (SBEM). Under UK Building Regulations Part L2, this has now to be carried out at the design stage and at completion to verify emissions compliance. Refer to Appendix of Approved Document L2, *Reporting evidence of compliance*.

Likewise with energy modelling: the MEP services designers can use software packages to model the energy performance of the completed building. This enables the designers and the customer to look at options to improve energy efficiency, if required.

In a BIM data environment this can be reviewed as part of the modelling process, using packages that interact with the BIM project model.

Figure 4.1

Code for Sustainable Homes

The Code for Sustainable Homes is similar to BREEAM, but applies to domestic buildings. Again there is a series of headings under which the project is assessed.

Refer to the BREEAM website for further information: www.breeam.org.

The CSH covers nine categories of sustainable design:

- Energy and CO_2 emissions (M)

- Water (M)

- Materials (M)

- Surface Water Run-off (M)

- Waste (M)

- Pollution

- Health and Well-being (M)

- Management

- Ecology.

There are mandatory performance requirements in six categories (denoted by an M above). All other performance requirements are flexible. It is possible to achieve an overall level of between zero and six, depending on the mandatory standards and proportion of flexible standards achieved.

4 Tools

Contractor's Proposals

Refer to Section 5 Procurement.

Deliverables Schedule

This relates to the Design–Procure–Construct Programme, Information Required Schedule, Resources Schedule and Responsibilities Matrix.

The Programme should define when information is required for particular activities.

The Deliverables Schedule lists the information deliverables, individual drawings, schedules, specifications, etc. It should also show the originator/owner of the deliverable, and can then link to a Resources Schedule, which drills down to the individual people producing the information. That can then be used as a tool to check that the process is being resourced adequately.

The Deliverables Schedule should also make a distinction between information required for procurement (i.e. to be sent out with tender enquiries for subcontract work packages) and that required for construction. (Information for construction also should be sub-categorised so that information for subcontractors to start CDP design, and information that can be constructed from directly, can be differentiated.)

Consequently this information can be used to calculate the design fees on a resource basis – the 'bottom-up' calculation method I referred to earlier.

The Deliverables Schedule forms the basis of the Information Required Schedule, combined with dates from the Programme.

DM Project Plan

Most businesses will produce a DM Project Plan. It will contain basic information about the project, key contacts, details and protocols for DM on the project.

A typical set of contents might be:

- Project directory, key companies and contacts
- Project description and details
- CDM requirements
- Protocols for issuing and circulating information (this could involve the use of a web-based collaboration tool)
- Protocols for approval of drawings, consultants and subcontractors
- Typical reports forms and frequency
- Meetings schedule/agendas
- Change-control procedure
- Design/procure/construct programme
- Information Required Schedule and Design Deliverables
- Design responsibility matrix.

The document should be updated as the project progresses and other parties join the project, such as additional subcontractors.

Table 4.1 Deliverables Requirements/Information Required Schedule.

Package	Package Lead Designer	Information for Procurement Date Required	Final Information for Construction Date Required (Contract Issue)
04 PILING *CDP Package* *Performance Specified/Sub Contractor Design* This should be a performance specification setting out the minimum requirements.	***Structural Eng***		
Method of testing to be confirmed by Engineer		16/01/2012	20/02/2012
PILING SCHEDULE This must detail the loadings that the piles are to be designed to accommodate.		16/01/2012	20/02/2012
PILE LAYOUT DRAWINGS These should be at 1:100 and show the proposed locations for the piles.		16/01/2012	20/02/2012
07 STRUCTURAL STEELWORK CDP Package	**Structural Eng**		
Sub Contractor Design Input *Specifications:*			
– Structural Steelwork		*23/01/2012*	*12/03/2012*
– Protective Coatings		*23/01/2012*	12/03/2012
– Fixings / connections		*23/01/2012*	12/03/2012
– Secondary Steelwork		*23/01/2012*	12/03/2012
Drawings:			
Steelwork layouts, plans, elevations 1:100		23/01/2012	12/03/2012
Sections through building 1:100		23/01/2012	12/03/2012
Key details on interfaces 1:20, 1:5		23/01/2012	12/03/2012
Additional Information (if appropriate)			
Holding-down bolt details		23/01/2012	12/03/2012
Purlins or cladding rails for other trades		23/01/2012	12/03/2012
Connections and bracing requirements		23/01/2012	12/03/2012
Fire protection / performance design requirements		23/01/2012	12/03/2012

For an Information Required Schedule, the deliverables could be listed more generically, or sometimes the actual drawing list is used to specify what is to be delivered by when. Normally the more detail the better, but it is a question of being appropriate to the size and complexity of the project.

DQI – Design Quality Indicator

Refer to Section 9: Quality.

Employer's Requirements

Refer to Section 5: Procurement.

Fire engineering

Again not exactly a tool, but certainly a technique that has added value to some projects I have been involved with over the years.

Every project will require a fire strategy, to demonstrate compliance with the Building Regulations Approved Document B and supporting fire regulations, such as BS9999: 2008 *Code of practice for fire safety in the design, management and use of buildings*. However, this is a *prescriptive* approach – i.e. there are rules, and the rules tell you how to comply.

Using a fire consultant it is possible to adopt a different design approach, which is based more on assessment of risk rather than prescriptive requirements.

As an example, on a Central London office development, our tender solution included the option of removing one of the stair/lift cores, as it was surplus to requirements from an escape perspective. We were also able to reduce the fire resistance of some structural elements. This saved cost and created additional lettable floor space as value added.

On an education scheme we were able to suggest the omission of the sprinkler system, and also reduce fire resistances to certain elements, again saving cost for the customer.

Information Required Schedule

This document will integrate with the *deliverables*, and the *design programme*.

Typically the IRS will be a table or spreadsheet, listing items of information required against dates for issue, usually to the main contractor. There could be drawings, schedules and specifications individually listed, or an indication of what's required, e.g. 'piling details'. However, the clearer and more detailed the IRS can be, the better the chance of reducing problems later on in the project.

Rather than just a long list of items, the Schedule should be broken down into the Works Packages for the project.

In addition it is also useful to split the IRS requirements related to the stage of design or the contract. Is the information required for procurement of packages, or for full construction?

This will affect the quality and completeness of the information required. Similarly, design-intent information required for commencement of subcontractor design need not necessarily be at full construction issue status, but will need to be complete to enable subcontract design to start effectively.

Interface matrix

While the consultant design team will design a project holistically, engaging all the disciplines and elements as they progress, contractor-based design currently is much more individual element- or system-based.

This means that there could be several subcontract works package designers engaged on the project at the same time – or perhaps not, as sometimes the vagaries of the procurement programme might mean that certain package designers will not be appointed until sometime later in the project programme.

This is an important concept to note – sometimes the procurement programme should be driven by the needs of the design process, rather than by the construction sequence. Major packages such as substructure, superstructure, external envelope, roof and MEP services ought to be completing their detailed design concurrently. It will make the process of interface management and coordination easier.

The plethora of packages results in the need for an interface tracking mechanism. The degree of success in managing the interfaces will determine the success of the project. Poorly managed interfaces can lead to delays in design and construction, as well as abortive work and needless redesign. This can be avoided.

In its simplest form, the Interface matrix will consist of a spreadsheet correlating which packages interface with which. Some matrices allow you to append more information, so as to describe in more detail the issues and key points. Other forms of matrix are like a package schedule, listing each package, the packages interfaced with and key issues to be resolved or checked.

In package design management there are two strands to the design works – firstly the main bulk of the package that is central to its scope, the core of the package if you like; and secondly the boundaries of the package and how it touches and joins up with the packages surrounding it, its interdependencies. Managing both strands effectively is critical to successful design management.

Constant review, through the approvals process and workshop reviews, is required. Frequently it is in the interfaces that additional cost is incurred, because the interfaces were not considered early enough and not allowed for in the Cost Plan.

In many cases it is easier and more productive to get everyone around the table to thrash out the details and agree how the interfaces should work. Understanding the sequence of assembly and construction plays a large part in dictating how everything comes together. One package will be sequentially dependent on others in determining how the interfaces are successfully constructed on site.

Interface Matrix

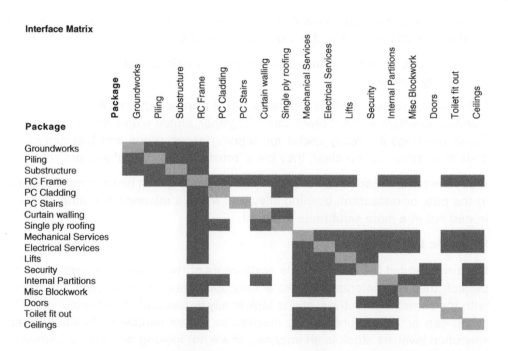

Whilst a matrix identifying the relationships between Works Packages is useful, it will depend on the level of detailed information about the interfaces and how this can provide the agenda for co-ordination workshops.
So for Ceilings/RC Frame/MEP Services
Items to be considered
    Fixings to soffit
    Hanger layouts
    Perimeter edge details
    Services in ceiling voids
    Services penerations
    Lighting layout
    Vent grilles
    Other service penetrations
    Fire protection
    And so on...

Figure 4.2

4 Tools

Matrix of Package Responsibilities

This is another matrix or table, listing all of the work packages on the project. It describes who is the lead designer for each package, and other designers who have an input or supporting role.

Usually this breaks down into Architectural, Structural and MEP packages.

Meetings

At the outset of any project it is important to have a communication strategy to ensure the smooth running of the design and information flows.

There are a number of key meetings that need to take place:

Project Launch meeting

- Possibly the most important meeting of all on the project, this is when the whole team meets together, perhaps for the first time.

- Use an external facilitator or someone not involved in the project. Meet at a neutral venue if possible, e.g. a hotel or conference centre.

- Get everyone – the customer, designers, contractor's team and key supply chain – to explain what they do, their roles and their aspirations for the project.

- Have a few break-out sessions to consider key aspects of the project or other issues. This is as much about building relationships as sharing information.

- If possible, try to get at the key challenges or risks of the project and discuss the way forward.

This meeting will set the tone for the project. It is important that a common understanding develops and the team comes together, united with a common purpose. These meetings are really useful for aligning expectations and just getting the lines of communication clear; they are a 'must' for the start of any project.

Also remember the social side. Sometimes a 'meeting' can be even more effective in the pub, or restaurant, bowling alley, etc. Many a misunderstanding has been ironed out in a more salubrious location!

Strategic Leadership meetings

Usually attended by the principals or directors of the businesses involved in the project, these are intended to take a 'helicopter view' of the project and to deal with the strategic objectives, and to look at any issues and identify solutions. This group can act as an unblocking mechanism, when people at detailed project-execution level are stuck in an impasse, or are not looking beyond the immediate details.

Design Progress meetings

They are just that: to review progress, actions, information required, compliance issues, etc. The elementary mistake that people can make is to mix progress review with solving technical issues. The result is that the meeting loses focus and becomes very wearing. While doing my Part III year, I worked with a job architect on a medium-sized, reasonably complex project, and the site meeting lasted most of the day. We discussed everything under the sun in the utmost detail. Losing the will to live didn't begin to describe it! – and writing the minutes was a nightmare!

Table 4.2 Tracking schedules come in various forms, and are known by various names, but the principles are essentially the same: Establishing WHAT has to be done by WHOM and by WHEN.

Works Package Responsibilities (Example 1)

Package No.	Trade	Lead Designer	Input by Others
01	Piling	Structural Engineer	
02	Curtain walling	Architect	Structural supports/frame details by Structural Engineer
03	Aluminium Windows/ External Doors	Architect	Security installation interface
04	Roof cladding and gutters	Architect	Syphonic RW system outlet/gutter capacities by MEP Services Engineer
05	External wall cladding	Architect	Structural frame interface/windposts/rails
06	Fire stopping	Architect	In-trunking fire insulation by MEP Services Engineer
07	Mechanical (may be split into more packages)	MEP Services Engineer	Specialist subcontractors. Lead Mechanical subcontractor co-ordinates and draws minor builder's work for all services and drainage
08	Electrical (may be split into more packages	Services Engineer	Specialist subcontractors.
09	Structural steelwork	Structural Engineer	Architectural steel and finishes (including intumescent paint by architect). Connections by subcontractor.
10	Structural slabs	Structural Engineer	Penetrations by Services Engineer. Coordinate MEP requirements.
11	Lift Installations	Services Engineer	Finishes by architect Guides ad fixings co-ordinated with Structural Engineer Routes and plant areas for hydraulics.
Etc	*Etc*	*Etc*	*Etc*

Design Package Responsibilities (Example 2)

Ref	Design Package	Consultant
1.	Architecture/Space Planning	Architect
2.	Structural Engineering	Structural Engineer
3.	Civil Engineering	Structural Engineer
4.	M & E Installations	Services Engineer
5.	Lift Installations	Services Engineer
6.	Information Technology	IT Consultant
7.	Public Health (below ground)	Structural Engineer
8.	Public Health (above ground)	Services Engineer
9.	Acoustics	Acoustic Consultant
10.	Fire Engineer	Fire Consultant
11.	Hard Landscaping	Architect
12.	Soft Landscaping	Landscape Architect
13.	Interior Design	Architect
14.	Catering	Specialist Catering Consultant
15.	Planning Supervisor (CDM Regulations)	CDMc

(Continued)

4 Tools

Table 4.2 (*Continued*)

Design/Package Design Responsibilities (Example 3)

No.	Package	Full Design by consultants	CDP	Consultant Info/comments
01	Demolition	Yes		
02	Groundworks	Yes		
03	Piling		Yes	SEng – loads, layout, diameters
04	Structural Steelwork		Yes	SEng – loads and layout
05	M & E installation		Yes	M+E – performance spec
06	Windows, Curtain Wall and Doors		Yes	Arch – performance spec and design intent, spec, details
07	Reinforced Concrete Frame	Yes		Includes rebar schedules
08	Lifts		Yes	M+E – performance spec Arch – finishes Seng – fixings/guides loadings
09	Metal Roofing		Yes	Arch – performance spec and design intent drawings
10	Wall Cladding	Yes		Assumes render/lightweight framing solution
11	Brickwork and Blockwork	Yes		Advice on fixings/windposts

On another project, a large commercial one in London, I led the design review meetings. The first hour, in the morning, was the formal progress review. We reviewed the design programme, information flows, etc., recorded required actions and so on. After the progress review we moved into a more informal Technical Review, which lasted the rest of the day, where we looked at all the current technical issues and turned the meeting into a design workshop. As a format, this worked very successfully and kept everyone focused. It also enabled different people involved in the project to attend the relevant part of the meeting during the day, maximising efficiency of use of their time. We would look at elements and packages of the design on a rolling basis, or as particular issues were arising.

Workshops/Interface meetings

These have been discussed under Interfaces above. For package review meetings it is essential that the DM is working closely with the commercial and building teams, to ensure that cost, buildability and programme are adequately covered. See also DM and Cost Management in Section 2.

Post-completion feedback workshops/meetings

Next to the Project Launch this could be the most important meeting on the project. It is invaluable to review performance of the project after handover. It could take in a number of aspects, such as:

- Team performance

- Effectiveness of communication

- Delivery process

- Quality

- Lessons to be learned and carried forward. This is particularly relevant where the whole team might work together again on another project, such as on a Framework, for instance. Effective communication of lessons learned is critical for achieving continuous improvement and innovation.

4 Tools

At the end of a project it is useful for the team to consider whether the customer actually obtained what they originally wanted. Did the team meet the brief?

Post-occupancy review is a related but distinct activity, covered by BSRIA and their Soft Landings process.

Models, virtual, physical

A picture paints a thousand words, and the same could be said of a model, whether that is a physical model or something from virtual reality using computer visualisation.

It takes several years to become really adept at reading 2D drawings and appreciating what they mean in three dimensions. Even then it is possible to miss something vital. A visual, a model, a rough prototype even, can help to illustrate a project or an interface problem that will help you arrive at a solution.

For a tender on a commercial project I made a model from card, balsa wood, polystyrene and anything else I could get my hands on. The topography of the site and the construction sequence were complicated, owing to existing protected features of the site and the architect's design. The model was composed of the contours of the site, and the floor plates of the building mounted on card. It enabled us to explain the construction sequence and logistics issues graphically at the interview. It also got people talking and discussing the issues, which loosened up the interview for us. Sometimes it is not always about technology!

Peer reviews

These are about getting perspective – a different way of looking at things; sometimes we can get so close to details and problems that we lose objectivity and the ability to look at situations differently to see the way through.

A quick review by a trusted colleague can provide fresh insight and solutions to seemingly intractable problems. Or just simply talking it over with someone, outlining the difficulties, potential solutions and weighing the courses of action, can help.

This is where working in a broader DM network can be of help. Where there are several DMers in the same company, why not informally agree to look over each other's projects occasionally? Even if there is no formal organisation or forum for your business, why not set one up? This could even form part of your CPD requirement, to review projects and discuss ideas and best practice. It can only serve to improve your delivery of DM, and that of your business – everybody wins.

For me this has been a huge plus point for the Design Managers' forum on LinkedIn (www.linkedin.com). DMers from all sorts of backgrounds and situations can share knowledge, ask questions and get help.

> Whatever issue you have on your project *now*, I guarantee that there is at least one person in the world who has already had the same issue, solved it and moved on to the next problem! So this is one of the reasons why it is good to connect and network with other DMers – to share knowledge, experience, problems, solutions and ideas.

4 Tools

Planning (Development Control)

Planning permission can be either 'outline' or 'full'. Outline is more about establishing the principles of a proposed development and is based on minimal information. Full planning naturally is based on full details of the scheme, usually design information approaching RIBA Stage D.

The approval will come with a schedule of conditions and 'reserved matters', which must be discharged by the submission of further information from the design team and supply chain. The notes on listed buildings below cover this process.

Listed buildings

Under UK planning legislation, buildings of architectural and historic interest are 'listed'. Grading varies for Grade II, Grade II*, or Grade I. The latter are buildings of national historical interest, e.g. St Paul's Cathedral, the Palace of Westminster, etc. Under the listing details for the project, there will usually be a citation stating the particular features or reasons for listing.

Any work carried out on a listed building or its curtilage must have permission prior to commencement. The application is just the same as applying for planning permission, but it is a separate process, and a separate approval is issued.

A building that isn't listed but is situated in a conservation area is subject to conservation-area consent, which works in a similar way.

Whether it is planning, listed, or conservation consent, an approval when issued will have conditions attached to it in a schedule. This will contain a number of conditions to be discharged by the submission of further information. In most cases the information is to be supplied by the design team, but sometimes by particular subcontractors or the client.

The process of discharging conditions must be managed to ensure that conditions are indeed discharged and design approved before work is ordered or installed. Otherwise the finished work will be at risk, and changes may be required by the Planning or Conservation Officer, or perhaps by English Heritage.

Several years ago I worked on the refurbishment of a listed bank building. The banking screen made by a specialist shopfitting company had been installed before consent was obtained. The English Heritage officer concerned would not approve the design. In the end the installed screen was scrapped and the design changed and remanufactured. An extreme example, perhaps, but a salutary lesson in making sure that you discuss details and apply for approvals at the right time.

The simplest method of dealing with this is to compose a schedule listing:

* The conditions

* Information required to discharge

* Who is the owner or lead on the condition?

* Who will produce the information?

* When will the information be completed, ready for submission?

The schedule should be reviewed regularly at design-team meetings, until completely discharged.

Note – While it is possible to start work on site in certain circumstances without Building Regulations approval using a Building Notice, starting without Listed /

Table 4.3 Compliance Tracker.

Planning Approval Conditions

Condition No.	Condition	Lead	Support	Start on site	Approval reqd by	Submit by
12	Provide details of replacement windows including elevational details at 1:50, with detail drawings at 1:10, with a sample window, that incorporates section sizes, glazing details, finishes and glass type. Details to be submitted and approved before window installation starts on site.	Architect	Window Subcontractor	11/06/2012	25/05/2012	02/03/2012

Where information is required only from the design team, this is relatively straightforward. However, if information is required from a subcontractor as above, then the procurement programme needs to take account of the programme timings required to discharge conditions.

Planning consent can have dire consequences (and is entirely different!), leaving the employer open to legal action by the local council. In extreme cases, enforcement action by the council can lead to removal or demolition of work carried out without permission, all costs being borne by the employer or building owner.

I have several times come across people who have not understood the legal differences between, Planning, Listed and Building Regulations legislation, with attendant implications for their project. (Remember that planning is about the principle of whether a development can take place. Building Regulations are about the technical compliance of the proposed building.)

So make sure you understand the main requirements, and the submission processes for each. Local authority departments are usually very helpful in providing guidance and information.

Programmes

We have already discussed the intrinsic connectedness of DM. In programming this is where it really does come together. Design cannot be programmed in isolation from the other phases of the project, so design must be considered in conjunction with the procurement and construction stages if you are to succeed.

In terms of integrated design–procure–construct programmes there are a number of characteristics:

- Firstly, the strategic milestones of the project. Formation of contract, start on site, phased completions, practical completion, handover, etc

- Secondly design activities by discipline e.g. architect, structures, MEP services, etc, including design-stage milestones, and reviews for RIBA Stage D, E, F, etc

- This should be followed by the works packages, showing for each package: design information preparation, preparation of sub-contract tender, documents, tender period, returns, contract letting, subcontractor design, approval periods, manufacture and installation periods. Included here could also be milestones for factory visits, samples, prototypes and testing

4 Tools

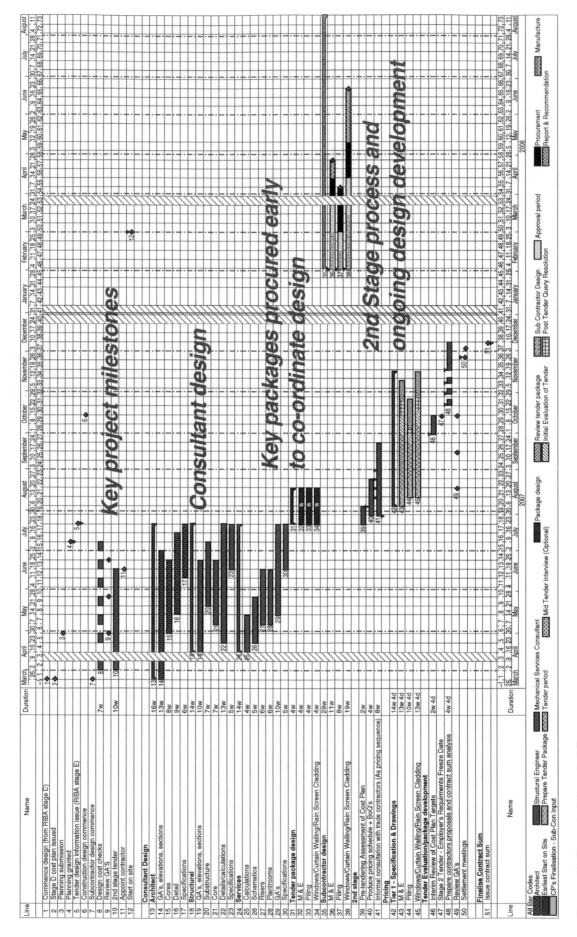

Figure 4.3 Example Integrated Programme.

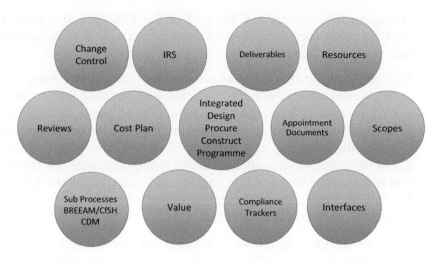

Figure 4.4 The DM Toolbox.

In working with the consultant designers preparing design-intent information for package tenders, it is advisable to agree reviews of the information which would occur, say, a few weeks before it is issued for tender.

These activities then link through to the construction programme.

- In addition, other activities can be overlaid on the programme, such as strategic leadership meetings, progress and design reviews and any other key events

Assuming the programme is produced in one of the main planning packages such as Primavera or Asta, then it is possible to produce detailed programmes for design, procurement, stages of construction, and each of the packages together with 'rolled-up' summary programmes, for ease of use.

Remember that the programme does not stand in isolation. Each of the bars represents an activity. Each of those activities will have outputs, which should be defined on your Deliverables Schedule and the IRS. Your matrix of responsibilities defines who leads and inputs on each package. The Contractor's Proposals define the scope, content, quality, cost and timescale.

Agreement of the Design Programme and Deliverables can be a complex activity in itself, but it is worth taking some time and getting it right at and from the outset. Consultant designers will be keen to see this resolved, as any commitments they make (or have already made) will need to fit within their agreed fees and resources. My experience is to try to have an open, honest conversation round the table, perhaps with a session or two at the pub, restaurant, etc.!

If everyone understands each other's needs, problems, expectations and can continue the dialogue, then there is every possibility that you will arrive at an equitable solution and programme. As Design Manager you stand in the middle of all these conflicting demands; if you can pull them together and maintain relationships in the process, then you are well on the way to success.

It will take a few iterations to get the activities and sequence adjusted. Inevitably it will be a compromise between allowing enough time for design and maintaining the construction milestones. There has to be a process of give and take between design, procurement, package and construction activities, to get the balance right. Quite often, working back from construction, you discover the design should have been completed last week. Working around this takes time, and usually there will be overlap to some degree between design and construction.

4 Tools

Managing this overlap is a key part of the DMer's role – particularly making package design decisions early on in the programme that do not create issues for later design packages.

Also remember that sometimes procurement should be driven by design rather than by construction. A project planner will typically lay out the procurement strings to serve the construction sequence. However, the design activities of several packages, such as structure, cladding and MEP services, need to happen concurrently, enabling co-ordination of details and BWIC (Builder's Work in Connection) to be managed more easily. It might be that once the design information has reached Status A, the package will not start on site for several months. No matter – you can keep the completed, fully co-ordinated design information in your desk drawer until it is needed!

There are design programming packages out there, of which the most notable is AdePT. This is a combination of process and software, compatible with planning packages and embodying lean principles. Appendix B by Paul Waskett and Andrew Newton explains it in a little more detail and also discusses the interface with a BIM environment.

A final word of caution here: project planners frequently produce their 2000 activity line programmes and proudly present them to the team. That's great so far as it goes, but creating a monster programme is one thing; monitoring, maintaining and updating it is another! I once worked with someone whose particular penchant was to list every drawing and activity on the programme, together with issue and approval periods and milestones. This created a huge, high-maintenance beast that quickly fell into disuse, simply because it was out of date as soon as it was produced.

Remember to create tools that are useful and accessible, in manageable forms.

Reports

Having created various processes, programmes and schedules for your project, you will need to establish some reporting mechanisms to see how the project is progressing, but just as importantly to flag up potential issues. My view is that reports need to be documents that are useful – not created simply to sit in a file or on a hard drive.

So WHY do you want a report by a particular business or organization?

WHAT will do you with it?

WHEN do you want the report? WHY at this time?

Consider these questions; the answers will vary according to your project, size, complexity and timescale. On one scheme we went from outline design to agreeing the contract in six weeks in a kind of crash design/costing programme exercise. We just met every week to review progress; the meeting minutes were the reports.

On larger projects, designers and subcontractors submitted reports every two weeks or monthly, depending on when the review meetings were taking place.

Subjects included:

- CDM/Safety

- Activities completed in the last reporting period

- Activities planned for the next reporting period

- Outstanding activities

- Scope/design changes

- Costs/fees

- Resources

- Information/decisions needed causing delay

- Potential issues.

The report itself is not so much the issue – it is more, WHAT are you going to do with it? The actions you take arising from the report information that you receive determine how successfully your process is working, and the validity of the reporting process. Producing reports that just sit in a file and do not lead to action are a waste of everyone's time.

Resources schedules

As discussed under deliverables, these are linked with resources. It is a useful technique to understand from your designers who exactly is doing what. Put names against drawings, check timescales, and look at this in relation to your design programme, the resources allocated and the fees agreed. Do these all add up? Are there enough people of sufficient quality and experience to produce the information to meet the programme? How does this stack up against the fees agreed? Does it make sense?

Quite quickly you should be able to make a judgement on whether there is a problem or not, and also what to do about it. (More fees? More resources? Get a supporting consultant? More time? Reschedule the programme?) The strength of relationships with your designers will be critical here in getting people talking and opening up about the real issues.

Risk assessments/schedules

Most businesses have risk-management processes in place. These will cover project risks across design, commercial and construction, together with actions to eliminate, reduce or mitigate. In a tendering situation cost allowances will be allocated against project risks considered to be significant and these costs included in the tender offer.

Health, safety and the environment will feature heavily in these processes, and rightly so.

Regarding design and DM, designers under the CDM regulations have a duty to review their design and produce Design Risk Assessments. Run-of-the-mill issues that a competent contractor would normally deal with can be discounted. However the designer has a duty to assess risk for any unusual features such as particular elements that need to be constructed out of normal sequence, or special requirements for temporary support etc.

This is covered in more detail in Section 2: Process. Remember that the DRAs are live documents. As the design develops, the DRAs should be updated. DRAs should be issued as part of the pre-construction information issued with the tender information. The CDM ACOP refers to this.

Resources Schedule

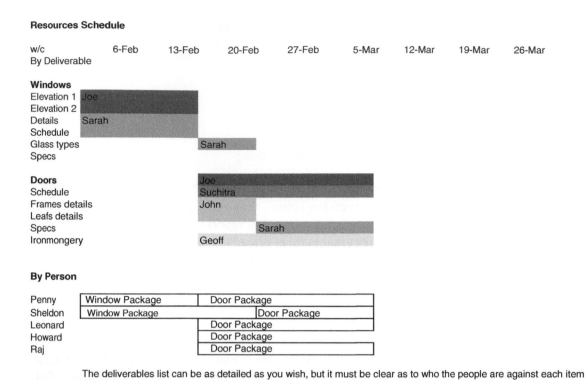

The deliverables list can be as detailed as you wish, but it must be clear as to who the people are against each item or groups of items. In auditing the resources you can then easily check the competence of the people against the deliverables. Cross check by mapping the people against the timeline.

Naturally dates and requirements integrate with the Design Procure Construct programme and the Information Required Schedule.

Figure 4.5 Example Resources Schedule.

Samples/Benchmarks/Prototypes/Mock-ups

As for animations above, a picture paints a thousand words – but if you can reach out, touch it, pick it up (if possible) and look at it from all angles, then that is much better than looking at 2D drawings and trying to imagine the 3D version.

The use of samples and prototypes can help the team to visualise exactly what is required. A mock-up of a particular detail can help resolve buildability or sequencing issues, which can be difficult to do with just drawings.

Discuss this with the team, review the package content and interfaces. Is there anything that would be assisted by obtaining samples, having an off-site mock-up made, etc.?

Another technique is to designate a section of the finished built work as the control sample or benchmark for the rest. Building and snagging proceed as normal but once the area or element is agreed, then everyone knows the standard to be achieved – just go and look!

Programme samples/mock-ups etc. into the design and package programmes. Make sure that their availability fits in early enough with the overall timetable to be useful. This may mean procuring a mock-up ahead of the main procurement sequence, but the information derived from having the mock-up may justify that approach. Your call – discuss the benefits with the team.

Schedules of services

These are simply WHAT people are going to do. These schedules are included in the appointment documents, or package contracts, and they need to be as clear and definitive as possible. Together with deliverables schedules, programmes and matrices of responsibility they define the scope of the consultant's or package contractor's involvement.

Scope documents

These are particularly related to Works Packages, a pro forma setting the content of the package, and referencing any design information such as particular drawings and specifications. It is useful to note any key interfaces with other packages, risks in design or delivery that require consideration, and any other relevant information.

Programme information can be included here as well. The scope document is used as a summary for that package. It cross-relates to the Interface Matrix. The scope can be used as a checking mechanism, to review package content and interfaces.

Using scope documents at the outset can be a useful way of getting a grasp of the whole project picture, understanding the package breakdown, key interfaces and design requirements. These can provide a platform for developing further detail.

Simulations

As we move into BIM environments, these will become much more common. It is possible to simulate carbon and energy use, people movement for queuing, or vertical movement, lifts, escalators, fire escape and fire spread, and so on.

These are useful design tools that assist in establishing the design with more certainty.

Stage reviews

However you decide key stages for your project, at the end of each stage make sure you have a formal stage review. Designers produce design reports at the end of RIBA stages. These can be a useful way of recording all of the work of the stage.

However, you have to question whether all the activities have been completed for the stage. Are the outputs as required? Is there anything outstanding?

Also, take time to review the next stage before commencing the activities and proposed outputs to make sure all your team are clear on what's required from them.

In some cases when reviewing a 'completed' stage it will become apparent that the stage isn't complete. So the decision will need to be taken to extend the timescale or carry over outstanding issues to the next stage.

Factors of cost, and pressure of programme, will decide the outcome. But whatever is decided the principles remain the same – maintaining clarity of deliverables, timing and any other requirements.

Status of information

Information is used all the way through a project. The deliverable should be defined, but to what use can it be put? How much can it be relied upon?

4 Tools

Generally information is issued with a status such as:

- Preliminary
- For Information
- For Construction
- For Approval.

Information issues for approval will be given one of three status ratings:

A – Approved (no comments) – Proceed to manufacture/construction

B – Approved with comments – Proceed to manufacture/construction with the comments incorporated

C – Not approved – Redraw/resubmit.

In approving information for manufacture or construction, remember that the information must be reviewed not only from a CDM/safety perspective, but design and compliance, commercial, programme, buildability as well.

Teams

Sometimes, owing to the size of the project, the DM function will be carried out by a team of Design Managers. The workload of the team might be split by design stage, discipline or subcontractor package breakdown, or perhaps by other ways of dividing up the workload.

However, in organising the DM team, the same basic rules apply as for doing DM, so:

- Make sure each role is clearly defined and scoped out.
- Understand the interfaces between roles, as items can be easily missed! (Remember the poem? *'There was an important job to be done and Everybody was sure that Somebody would do it. Anybody could have done it, but Nobody did it!'*)
- Ensure regular and clear communication.
- Hold regular review meetings, to maintain alignment as a team.
- Have clear objectives on time, cost and deliverables (outputs).

Tests

Elements of the construction will need to be tested. This might be because of requirements of the specifications, for example testing of cladding; or for statutory reasons, for example acoustics or pressure testing for Building Regulations compliance.

Understand the requirements for your project. Ensure that the tests are programmed in correctly to your delivery plan, allowing time for design, manufacture and testing if required and any remedial action required as a result.

Off-site tests will require a test facility, and they will need a reasonable lead-in to accommodate your project within the timescale that you need.

Tolerance/Movement Schedule

Buildings move under the forces of nature. Beams and floors deflect under loadings. Sunlight on one elevation will cause materials to expand, while a hard frost on the other elevation causes contraction. Materials are manufactured or constructed within certain tolerance limits (or variations). A number of the same type of bricks could vary by several millimetres in length, for instance.

The tolerances and movement that can be tolerated by a curtain-walling panel are quite different from those of a reinforced concrete structure or an adjacent panel of brickwork. Understanding this is crucial to how elements and components fit together. Similar considerations extend to how elements or components are fixed to each other, so in this way we have deflection heads for partitions or mastic/gasket joints between panels or elements, which remain watertight while accommodating substantial movements.

Therefore the movements or tolerances required for the various elements need to be known to understand how the building fits together. Very often these will be the interfaces between packages of subcontract design. On a recent project the concrete frame deflections were greater than could be tolerated by some large, glass sliding doors – the solution was to strengthen the slabs locally to reduce the deflections to suit.

A statement setting out the designed movements and criteria is a useful document at the start of the project, as these movements can then be allowed for as the design progresses, and included in package documents. Again, this is a living document and can change as the project progresses. Make sure it reflects the current design and the tolerances/movements required.

You Yourself

We look at people in Section 7, but you bring yourself to the process. What is your experience? What are your skills? Your way of thinking and problem solving, even? Who are you? What gifts do you bring as a person to this project – the 'softer' skills of how you interact with people, build and maintain relationships and communicate? The building industry is a very macho, testosterone-charged industry, which is changing – thank goodness – albeit slowly. But even now it is not usual to think about just the basics of how human beings come together and design and construct buildings. We need to, though, as any successful project is built by a team that works well together – and surprisingly that team is made of people, not computers or chess pieces! So get to know yourself, and to know what makes people tick. We'll look at these matters in more detail later.

And finally for this section

We have considered some tools that the DMer will use to control the project.

As you have seen, I have mainly considered these from a contractor's perspective. However, whichever side of the table you are sitting on, you will be dealing with the same sorts of issues and tools and so these should be familiar.

Always remember to question; take a step back occasionally and coolly review DM activities, and only use tools and processes that are appropriate, focused, and that your team will actually use.

In the next section, Procurement, we take a look at the main procurement options for projects in the UK and key aspects of DM.

5

Procurement

The CIOB Design Manager's Handbook

'Procurement is defined as the acquisition or obtaining of goods, works, or services, through a strategic process of:

- *identifying the need,*

- *assessing the supply chain,*

- *promoting the opportunity*

- *and selecting the final supplier.'*

– Simon Matthews

Introduction

In this section we will look at the main routes for procurement of projects. Globally these might come under slightly different guises or names, but the principles remain the same, particularly around the area of contractor involvement and responsibility.

The trend for several years has been to move towards contractor-led teams, both in the public and private sectors. This touches on the theme of the consumer contractor I mentioned earlier. Those that carry the most risk in project delivery will probably be best placed to lead the process and ensure success. This is a principle that may well transfer into BIM-enabled project environments, as we shall discuss later.

I am not going to dwell on the different contract forms such as PPC, JCT, NEC, etc., simply because there is plenty of guidance available already, and this isn't intended to be a discourse on contract law. We are just going to look at the over-arching principles involved. However, as a Design Manager it is well worth under-standing the various contract forms to some extent and how they affect the delivery process of your project and your role.

So in this section we will briefly look at procurement routes, and how these affect the Design Manager. In addition I'd like to discuss the Employer's Requirements and Contractor's Proposals, and also explore the sometimes sensitive subject of 'novation'.

The main procurement variants are:

- Traditional

- Traditional with CDP (Contractor's Designed Portion)

- Single-stage D+B (Design and Build)

The Design Manager's Handbook, First Edition. John Eynon.
© 2013 The Chartered Institute of Building. Published 2013 by Blackwell Publishing Ltd.

- Two-stage D+B
- Partnering/PPC (Project Partnering Contract)
- PFI/PPP (Private Finance Initiative/Public Private Partnership).

Traditional

In simple terms, this is probably the most straightforward format in that the customer employs the designers, they design the project, a contractor prices the design and once appointed, then goes on to build it. Essentially the contractor just works to the design information produced by the designers. Any DM that is taking place is happening within the design team, so the co-ordination of inputs will be led by the lead designer, also acting as DM – in this case probably the architect. However, on projects with a large structures or services input, then this could be the relevant engineering designer or specialist. Design information will be in the form of drawings, specifications and schedules, possibly with a Bill of Quantities. There might be a little 'unofficial' contractor-designed input for, say, windows or something similar, but it will be of a relatively low level of content.

It is arguable whether the contractor would need a DMer on a project like this. For best practice I would suggest a Delivery Design Manager could assist in making sure that the throughput of information is in the right form at the right time to suit the contractor's procurement and delivery processes. A Pre-construction Design Manager might be of assistance to the design team as they put the project information together. This input would be driven by the timing of the contractor's appointment and the agreed degree of involvement before work starts on site.

Sadly, although it is largely proven that early contractor involvement adds value, reduces risk and increases certainty around project delivery, some designers still prefer to keep the contractor at arm's length – that's their (and their customer's) loss, I'm afraid!

Traditional with CDP (Contractor's Designed Portion)

Similar to 'traditional' in principle, but the way some designers use this contract has more similarity to Design and Build!

Just the same as before, the customer employs the designers, who design the design. However, in this instance, they decide that they will need specialist input – to complete the design and manufacture and to assist the construction process – from the specialist subcontractor employed by the main contractor. This will be for specialist elements such as cladding, curtain walling, MEP services, roofing, partitions, etc. The designers will produce 'design intent' information, which sets out the parameters that the specialist subcontractor is to work to. This will be in the form of drawings and performance specifications that set out the requirements of the designers that the subcontractor must achieve.

There are a few things to note here.

Firstly, most designers almost always produce too much information. I think there is an opportunity with careful planning of the design programme to achieve more by doing less. More thought in framing the intent information, without being prescriptive about the possible solutions, provides the contractor and their team with the potential to add value through creative solutions involving their supply chain. Usually it is the expertise of the supply chain that designers are keen to leverage.

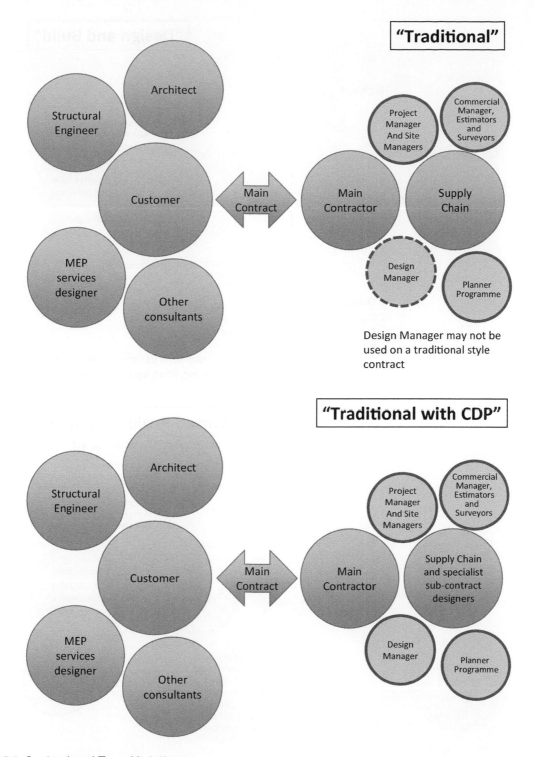

"Traditional"

Design Manager may not be used on a traditional style contract

"Traditional with CDP"

Figure 5.1 Contract and Team Variations.

Let's face it, there is no point in designing some cladding details to RIBA stage F2, only for the appointed cladding subcontractor to come on board and say at the first meeting, 'We don't do it like that! Our system is totally different.'

My understanding is that the lead designer is responsible for integrating and co-ordinating CDP design into the overall project design. The contractor is usually responsible for co-ordination between CDP design packages. In the past, accepted practice has been for a few packages to be CDP, such as MEP services, cladding/envelope solutions and a few others. This might amount to 30–40 per cent of

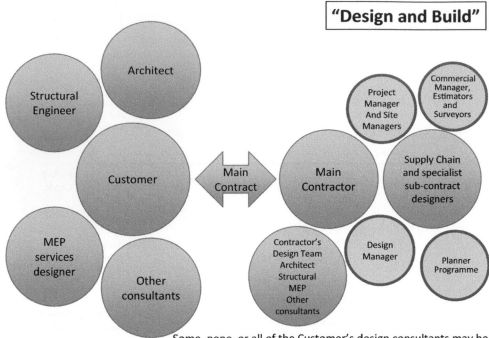

"Design and Build"

Some, none, or all of the Customer's design consultants may be
novated to the Main Contractor, at an agreed contractual stage.

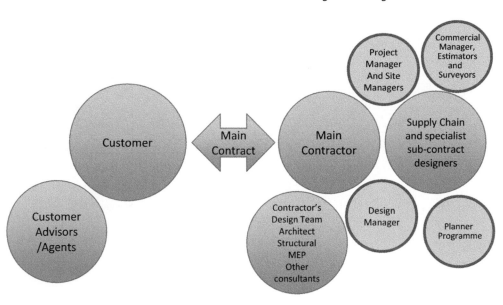

"Integrated Project Delivery"

Similar to Integrated Design Construct

Figure 5.1 *(Continued)*

contract value. (MEP services for a long time have been procured using detailed
design and installation by the MEP subcontractor(s). Rarely is full design carried
out by the consultant MEP designer, unless there are particular project requirements or circumstances.)

However, there has been a trend towards more and more of the project being let
as CDP, in some cases virtually 100 per cent, so this becomes effectively Design
and Build under another name. In some cases there is very little, if any, design

input by the subcontractor on some packages. This isn't anything to do with leveraging specialist design expertise, but more to do with transfer of risk – risk of delivery. The designers having made it clear what they want, they then stand to one side and 'dump' it on the contractor to deliver and carry the risk. Hence the term 'design and dump'.

There is nothing against doing this, so long as everyone understands what is happening and the implications. Sometimes the CDP design intent is so prescriptive that this closes out any real opportunities for the contractor's team to bring any value or innovation. Everyone loses out here – including the customer.

A word of caution about design-intent information; the designer still has a responsibility to produce design-intent information that is capable, with the subcontractor's input, of being translated into a workable solution within the project constraints and overall design. If the design intent is not capable of complying with Building Regulations, for instance, or working within the cost constraints, then this isn't the contractor's fault. Nor is it an acceptable risk for the contractor to bear. Used intelligently, the 'traditional with CDP' route can provide what it is designed to provide – supply-chain expertise to help the designers achieve the building they have promised to the customer.

Key to setting this up correctly is being clear about what you want the CDP to achieve. That will mean understanding the functional, technical, visual, environmental and any other requirements that must be achieved. Then be clear about how to communicate those requirements effectively and concisely, without repetition or deviation! Note that this can be implemented as a two-stage process for selection of the main contractor, similar to two-stage D+B (below).

Single stage D + B (Design and Build)

This procurement route has a few variations. However the overriding advantage for the customer is that it provides single-point responsibility for the project, and avoids having to deal with a plethora of parties who sometimes seem to be at war with each other rather than focusing on delivering for the customer. The adversarial, tribal image of the industry really isn't very helpful.

Firstly, a novated route – this is where the customer employs a design team to produce a solution. The solution is tendered and a main contractor appointed, to continue the design process and construct the project. The contractor might directly employ his own design team, or alternatively the contractor might 'novate' all or some of the customer's design team. This is where in effect the contractor takes over the customer's contract with the designers. The contractor might novate some of the customer's designers and also employ his own as well.

The design information that is tendered to select the main contractor can be extremely variable in quality and stage of completion. It can be anywhere from an RIBA stage D to stage F. This design information is termed 'the Employer's Requirements' – ERs. The contractor's tender, or response, is termed 'the Contractor's Proposals' – CPs. I may diverge with my commercial brethren as to defining ERs/CPs, but I'll discuss this in more detail below.

However, in practice, the ERs usually tend to be around RIBA stage D/E. In a 'design and dump' scenario the design has all but been completed (I have seen F1/F2 on some projects), statutory consents obtained and little opportunity left for contractor added value. This is the risk transference situation. The customer wants single-point responsibility, and so hands over the designers to the contractor, through novation.

The second scenario is an IPD/IDC route (Integrated Project Delivery or Integrated Design Construct). This is where the contractor comes on board at the start and leads the team. Project definition could be at an early stage and the main contract is between customer and contractor. Designers are appointed by the contractor. Advantages here are that a true value-management process can be adopted and early supply-chain involvement can be facilitated to aid the design process.

Novation

Novation is where the contractor takes the place of the customer in the designer's contract. After the contractor has been appointed, the contractor becomes the designer's customer!

Sometimes novation can work really well; however, at other times it can be like a shotgun marriage. In some tendering situations novation of the designers is a condition, and no choice is given. There are designers who have no desire to be employed by contractors at all. They feel they lose control of 'their design' and all that contractors do is cheapen the quality of the project in order to enhance their vastly inflated profit margins and make life easier – a rather outdated stereotype that undervalues the role of the contractor and supply chain in delivering designers' visions!

Perhaps when design-and-build was first off the blocks, then this might have been true, but I think it is a rather old-school, outdated view now. Through design-and-build procurement, customers and designers can access support from contractors and their supply chain that can add value, produce more risk-free design solutions and increase certainty around delivery of the project, earlier in the process. However, it could be said that I am biased!

Remember – whatever hat we are wearing at the time or whichever tribe we are a member of – we all just want to do a great job. Successful projects and happy clients can mean repeat business for all involved: everybody wins. I firmly believe that a form of IPD is the best-value solution for customers on most projects and the increasing use of BIM will enable this to only get better as a vehicle for project delivery.

Novation can be difficult, and can create tricky relationships. In the past I have had some designers who have continued their relationship with the customer as though nothing has changed, with scant recognition of who is now paying their fees (i.e. the contractor). Also, sometimes it is stipulated in the contract that the novated designer will continue to inspect the built work and report directly to the client.

This is to be resisted, as it creates split responsibilities and while designers will talk of so-called 'chinese walls' (i.e. someone else, separate within their business but not in the project team doing the reporting), this can rarely work effectively without creating conflicts of interest. Novation can bring together designers and contractors who in normal circumstances just wouldn't work together. In the words of the marriage service, it is 'not to be entered into lightly, irreverently or hastily'.

If all parties understand how their relationships should work post-novation, and their responsibilities, then the better it will be for all. As a rule a novated designer should not be in direct contact with the customer without the DMer being aware of the conversation and being able to input/direct as necessary. Customer changes, instructions, and any communications at all should all be directed though the contractor at all times, for 'the safety and comfort' of all concerned.

On a positive note, novation can be a very good experience, with designers and constructors learning from each other and going on to work together again in the future. This has happened in my experience more than a few times, so it is not all doom and gloom here.

If novation is on the agenda, it is well worth both intended parties getting to know each other. Certainly as a DMer I would want to understand the designer's approach to the project, attitude to D+B, the resources they have available, their experience and so on.

As a designer I would want to know how the contractor approaches Design Management, in agreeing programmes and deliverables, managing the process, amendments to appointment documents, selecting supply chain, and attitude to design quality and integrity.

Just like a marriage, if people understand each other as much as possible at the outset and align their expectations, then the relationship is more likely to be successful!

Two stage D + B

If a form of IPD is my first choice for providing customer value, then this would be my second – as long as certain criteria are complied with. In this procurement scenario again there are several variations.

The customer employs the design team. They take the design to a suitable stage, usually somewhere between RIBA stages D and E. The design status does not need to be much further than this, or the opportunity for creating value through contractor and supply-chain involvement will be reduced.

A first-stage tender is issued on this basis and a main contractor appointed, usually on the basis of a limited second-stage scope of service, for a fee, overheads and profit. Consideration is usually given in the selection process to the team, resources and service the contractor can provide, and feedback on the project cost plan.

Formation of the main contract is usually dependent on satisfactory completion of the second stage, developing the design to an acceptable level of detail, involving supply chain as required and carrying out enough market testing and tendering of subcontract works packages to enable satisfactory agreement of a contract sum.

Once the contractor has been appointed after the first stage, then the design is developed (in theory anyway) through the second stage with the input of the main contractor and potential subcontractors – particularly in terms of detailed constructability, logistics, consideration of alternatives and options, as well as tendering of key packages and including discussions with subcontractors. This is all carried out against the backdrop of validating the project cost plan and ensuring that the aim of satisfactory completion of the second stage is achieved to enable formation of the main contract.

Therefore, to include all these activities and to enable them to be completed to a reasonable degree of satisfactory detail, there needs to be a second-stage period of probably at least 10–12 weeks, and preferably longer. The second-stage programme needs to include a period for design development, to allow the contractor's team to get inside the design with the supply chain, agree some potential solutions and then tender these to subcontractors. Having obtained tender returns after, say, 3–4 weeks, these will then need review by the whole team before drawing together the final proposals and contract sum at the end of the second stage. Sixteen weeks is probably more realistic as a good-length period, allowing eight weeks for design development and review, four weeks for market testing and tendering, and a final four weeks for review and final proposals. Even this could be viewed as a little compressed; it depends on the stage of design, what degree of involvement is required and how quickly and realistically it can all be pulled together.

In my view, if a two-stage D+B tender has a significantly shorter period for the second stage, then either the customer and design team do not understand the implications of what they are doing, or they are not interested in any meaningful contractor input. A successful second-stage process will add real value to the project and help the customer and their team in achieving their objectives, as well as providing the confidence that their project can be delivered with certainty.

The Employer's Requirements

As discussed earlier, this is the statement of the customer's requirements that the design-and-build contractor and their team will tender against. The information released at this stage can be extremely variable in both quality and quantity.

At its simplest, it could consist of a statement of requirements, with a RIBA Stage D design plus an outline specification. On other projects I have experienced reams of information and drawings virtually at construction status, with specifications, Room Data Sheets, full structural and MEP design.

As we have considered before, adoption of a D+B procurement route can be more about transfer of risk rather than providing the opportunity for the contractor and their team to add value to the project. In this case the customer's team will have defined the requirement to the nth detail, ensuring that very little can be altered without difficulty – 'Design and Dump'.

Underlying it all is the assumption that ERs are actually what the customer wants. That may not be the case, and if we follow the value-management chain, then the whole team needs to understand the customer's brief. Compliance with the brief needs to be verified and confirmed at every stage.

The Contractor's Proposals

These are the contractor's response to the tender requirements. The format of the CPs will depend on the commercial strategy for the tender, so they could be very brief or very full, rather like the ERs! It will also depend on what is required under the tender instructions.

However, as a rule there will usually be some kind of design response. A formal audit of the tender issue design information is prudent, as it will highlight key risks, particularly in terms of buildability and statutory compliance.

Other areas of the CPs could include value-management ideas, perhaps increasing nett lettable area, or offering alternative specifications for certain elements or systems, which could provide savings or other benefits. It is useful if the tender assessment criteria are available, as this gives a steer to the contractor as to how to marshal their response accordingly.

The ERs/CPs conversation

Once the tender and CPs have been submitted, there begins a process or 'conversation' between the customer and his advisors/team and the contractor. In the CPs the contractor will have probably made a number of changes to the ERs and offered alternatives. These could be to add value, save cost or time, improve buildability or site delivery logistics or alternative supply-chain factors. Together with the tendered cost, time and any qualifications or clarifications, they will form the 'offer'. Once that has been amended to mutual agreement and formally accepted, it will then form the basis of the project contract.

It is a critical moment for the DMer.

It is the transition point, and you are now in delivery mode. The project parameters are set, and the envelope or boundaries within which you and your designers (if you have them) are now working are defined:

- Time (programme) (When?)

- Cost (contract sum) (How much?)

- Quality (both standard and content, function, sustainability etc) (What?)

As the Delivery DMer you need to know the agreed CPs inside out, and so do your designers and any design subcontractors. The time for developing design in terms of exploring ideas or alternatives is over, unless it is within the scope of the agreed CPs. The mode now is simply of making it work, closing down the alternatives, and getting the design information through manufacture into installation and construction, within the parameters agreed under the contract.

I say 'simply' because, to be blunt, some designers 'get it' and some do not. It can be hard work getting people to focus just on achieving a buildable solution within the cost and time boundaries. It can be the same with some subcontractors, who will gladly expand, enhance and embroider their works package in the hope of gaining 'extras'.

So the delivery end of DM is far from straightforward, but in controlling time, cost and the outputs, the agreed CPs are your guide. Remember, it is the agreed offer, and it is what everyone has signed up to and it defines the boundaries of delivery of the project – 'the project envelope'.

And finally for this section . . .

While I am not expecting you to be an expert on contract law, I do expect the professional DMer to understand the implications of your particular project procurement route, both on your role and also on the process that you are following.

You must understand the brief, the ERs and the CPs. As a delivery DMer, the agreed CPs provide the context, discipline and boundaries for your daily activity.

You, your team, designers and supply chain need to understand them too.

Value and innovation

The CIOB Design Manager's Handbook

'Not everything that counts can be counted and not everything that can be counted counts.'

– Albert Einstein

'Price is what you pay. Value is what you get.'

– Warren Buffet

'If a window of opportunity appears, don't pull down the shade.'

– Tom Peters

Introduction

Value can mean many things to different people. Something that is valuable to me may be worthless to you. However, the cost of that item can be easily established. As to its value, that's a different question.

'What is the value of Design Management?'

This is a question that recurs because some people remain to be convinced that having a DMer on a project provides value. Meanwhile they're concentrating on the resource and prelim costs, but not the effects or outcomes. Perhaps it is an equation that cannot be balanced to the bean counters' satisfaction? This will not satisfy some people – I make no apology. Life is about more than reducing everything to numbers!

If I enable you to avoid a potentially fatal accident on the motorway, what is the 'value' of that? We can perhaps quantify the costs involved, but the value of your life is something else!

Every day DMers prevent mistakes in design, procurement and construction, through communication, review and collaboration with their peers and teams. In addition they contribute to the overall quality of delivery and the end product, through improving the working process and also the end solutions.

Our commercial brethren have their work cut out keeping tabs on what is actually happening, let alone putting costs to things that didn't happen!

'What value can Design Management create?'

Instinct tells us that if a DMer on a project is helping the whole team to communicate and move information about design more effectively, whilst establishing

The Design Manager's Handbook, First Edition. John Eynon.
© 2013 The Chartered Institute of Building. Published 2013 by Blackwell Publishing Ltd.

reviews, workshops and meetings with relevant people, designers, commercial, constructors and supply chain at the right time, then that must make a difference. Taking this a step further through innovation and looking at value options, DM can be a force for the positive, catalysing opportunities as well as simply preventing what are effectively mistakes, whether by individuals or collectively through lack of efficient process or expertise.

In this section, with the help of Michael Graham of UK Value Management, we're going to do some thinking around the subject of value and innovation. Michael's paper 'Delivering Value – a guidance note for Design Managers' is included at Appendix D. We have also included some ideas for further reading and research in the bibliography.

What is value?

'Value is a measure of how well an organisation, project, or product satisfies stakeholders' objectives in relation to the resources consumed' – EN1325.

Ideally, the factors of quality, cost and customer satisfaction should control project decisions in a planned and considered way. Unfortunately this is rarely the case, as the typical knee-jerk value engineering exercise often kicks in when the cost plan or tender is over the budget limit.

As an observation, this will be happening because the design/cost dialogue hasn't worked effectively somewhere along the line. Perhaps the cost plan hasn't kept pace with the developing design, or the commercial team hasn't appreciated the real potential cost of what has been designed. The reality check arriving in the form of the tender return frequently means the contractor is the bearer of bad tidings, and the resultant VE exercise is rarely popular, can be painful and never recoups all of the planned savings anyway. Perhaps this contributes to contractors sometimes getting bad press on delivering quality?

In reality the project started to go off track further upstream, as expectations, cost and quality diverged and went their separate ways. This is another argument for adopting a model of Integrated Project Delivery for project procurement, as early integration of the whole team enables the customer, designers, commercial teams and constructors to develop the design with more certainty and confidence. The adoption of BIM will only serve to reinforce integrated working and closer collaboration in the long term.

Right from the start

Returning to the analogy of the journey as project process, if you make a slight error somewhere along the way, then once discovered it is straightforward enough to retrace your steps back to the last correct milestone and go the right way. However if from the start you go in the wrong direction, even slightly off course, and just keep going, then eventually you will be a long, long way from where you should be. It is then much more difficult to find a way back.

So in terms of real value, we need to go right back to the start of the project. We need to get under the skin of the customer, understanding their business (if they have one or are one), their processes, organisation, culture, ethos, drivers, etc.

What are their expectations for the project? What factors are causing them to take this action or intervention? What impact will the project have on their business or organisation? What will it deliver? What will it do for them and their business?

If we have answers to these sorts of question, then this will direct our input on the project. Also at this point it should become clearer as to what potential solutions could add value over and above what is already contained in the project concepts so far.

How often has it happened that the customer's situation and drivers haven't really been understood at the outset, with the result that at completion the customer has a completed project that doesn't represent everything they really wanted?

A better understanding of the customer, their objectives and drivers, right from the start, makes subsequent decision making a little easier. Any required decisions can be judged against this background and their relative merits measured and established.

The Value-Management (VM) framework

VM provides a framework to recognise, measure and improve value. The framework is about the fit with customer objectives and creating a value culture within the project context, so that every decision at every level aims to add value.

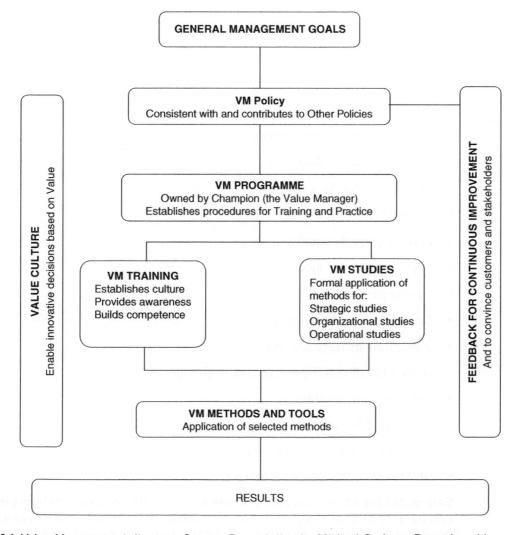

Figure 6.1 Value Management diagram. Source: Presentation by Michael Graham. Reproduced by permission of Michael Graham, UKValueManagement.

This framework enables organisations to innovate and to plan and deliver value for money. The standard **BS EN** 12973 illustrates the value-management framework in the context of corporate goals and the establishment of a value culture.

The five core value-management methods are:

- Function analysis – how things work – analysis of situations and problems
- Function cost
- Function performance specification – establishing objectives/outputs
- Value engineering/planning/analysis – optimising solutions/stimulating innovation
- Design to cost – exploration of cost, benefits, short-term/long-term.

Value management delivers:

- Innovation by focusing creativity
- Productivity by focusing on value
- Sustainability by incorporating all stakeholder views
- Consideration of environmental, community and financial inputs
- Customer satisfaction
- Improved business performance
- Innovation
- *Over the whole asset life cycle.*

VM methods are applied throughout the project lifecycle to:

- Analyse business requirements
- Make sustainable decisions
- Strengthen partnership teams
- Agree performance targets
- Prepare project specifications
- Improve value for money
- Save time
- Stimulate creative designs
- Streamline delivery process
- Deliver effective, efficient, economical projects
- Improve team building
- Achieve better solutions
- Increase customer satisfaction
- Create and add value – right first time.

Setting out the VM stall as above makes some significant claims in terms of what can be achieved on projects. There is convergence with both lean thinking and the use of BIM in delivering optioneered VFM solutions, efficiently, with an economy of resources that provides the customer with value over the whole asset life cycle.

DM + VM

Design Managers have a significant role in applying VM more effectively.

Frequently, Value Engineering is the most visible aspect of VM on a project – often applied too late in the process and leading to short-term solutions that are focused interventions and predominantly cost-cutting-based, with limited success.

In a wider context, the object of value management is to 'maximise the overall performance of the customer's organisation.' This is much broader than the usual project-limited view, and takes us back to 'Right from the start'.

Cost cutting in one element will have repercussions somewhere else, whether that's in the quality of the end solution, the life-cycle cost, carbon emissions or replacement cycles.

Rather than doing less, can we find a solution that is lower cost to build but delivers better performance, in terms of profit for the supply chain and ongoing benefits for the customer?

Improving value is about improving the relationship between function and cost over the life cycle of the asset. That could mean reducing cost without corresponding performance loss from the product, or doing something innovative at a different cost to achieve higher performance levels.

VM isn't just about having a few workshops, but it is the application of the various strands of the VM framework to enable project stakeholders to make sustainable decisions that aim to maximise the overall performance of the organisation – *right from the start!*

For the Design Manager, various aspects of the project could be examined in turn as follows:

- Value *in* design – achieving the output specification and delivering the end product that (we)/industry/customer can build/lease/occupy and make a healthy profit

- Value *of* design – business/customer focus, rather than project focus, consideration of our values, brand, ethos, and priorities for long-term business success

- Value management – approaches, standard processes, toolkit.

In one of our CIOB DM workshops we discussed the relationship between DM, VM and Innovation and developed a few thoughts that, while not providing all the answers, could give some pointers as to how DMers implement this thinking on their projects:

How should the DMer stimulate innovation and improve performance?

- The project environment needs to be conducive to innovation (speak freely, discuss ideas supportively, refer to 'The Rules' in Appendix G: Facilitating workshops).

- Define innovation in the project context – is it really wanted? (Everyone likes the concept of innovation, but they often prefer you to innovate somewhere else first, at someone else's expense!)

- Ignore the project boundaries and constraints.

- Go the 'extra mile' (this is going to take some effort: ideas are 10 per cent inspiration and 90 per cent perspiration!)

- Whole team 'on board' – everyone needs to be actively engaged.

- Facilitation will be needed – use someone from outside the project? Inspirational skills/abilities needed to pull out good ideas and best practice from all stake-holders, designers, consultants, contractors and supply chain.

- Explore the perception of risk (Edward de Bono's 'Six Hats' method is a balanced way to examine ideas while maintaining perspective on advantages, disadvantages, challenges and risks; see Appendix G: Facilitating workshops)

- Support – ideas, and the individuals putting them forward, initially need some encouragement, rather than negativity; refer to The Rules.

- Allow time to 'cool off' /reflect within the programme.

- DMer needs to allow the designer(s) time to 'design' within the programme, to consider ideas and explore options.

- Collaborative working environment required, not adversarial as the latter saps energy, inspiration and the culture of generating insights.

- The culture to allow innovation/positive encouragement comes from the top – supportive business leaders/sponsors and project leadership.

- Knowledge base – DMers and members of the team develop solutions in iterations over time; with their experience over several projects, this knowledge can remain in silos or with the individuals. How can we share this knowledge to be more effective?

- DMers need to encourage and support better communication to allow creative culture.

- Innovation needs recognition to develop; this recognition and the benefits need to be

  - tangible

  - measured.

In developing an approach to innovation, the ground conditions must provide the right nutrients. Attitudes, culture and support mechanisms must be in place to ensure the innovation mindset can flourish, with the illusion of being 'safe'. Seth Godin in his book *Poke the Box* explores this in some detail and is worth a read (see the bibliography).

> *'Innovation is not safe. You'll fail. Perhaps badly.*
> *Now that we've got that out of the way, what are you going to do about it? Hide?*
> *Crouch in a corner and work as hard as you can to fit in?*
> *That's not safe, either.*
> *Might as well do something that matters instead.'*
> – Seth Godin, *Poke the Box*, p. 82.

What do DMers need in order to deliver better performance through more effective practice?

- Practical competencies and skills/experience

- Create the right environment/right culture

- Plan time to think/cooling-off period/reflection time during design/stand back and review

- Inspiration

- Funding/investment/support in shadowing/mentoring

- Co-operation/support of peers and leadership

- Effective systems/methodologies to collate ideas and implementation plans to turn creative idea(s) into reality

- Supply chain in place – R&D/new products

- Infrastructure in place to allow support group(s) (at pre-tender stage)

- Statement of need/stakeholder buy-in, business case/brief

- Recognise their worth.

Measures of success

- Project delivered on time and error-/defect-free, i.e. right first time and within budget

- View Requests for Information (RFIs) as Non-Conformance Reports (NCRs), i.e. if there are none, then the design conformed, was well managed and well thought through, and delivered correctly

- Few RFIs/NCRs would equate to a better design, i.e. are sufficiently precise and compliant

- Degree of compliance with the output quality specification

- DMers need to appreciate how the output/project will be measured/assessed at the end.

- How well does the delivered project provide the desired impacts on the customer and on their business, and meet the customer's success criteria as established at the outset? (Remember, right from the start . . .)

And finally for this section . . .

A value-management process applied correctly from inception of the project can be beneficial to all the stakeholders and the outcome.

All too often VM is seen as a very limited intervention, usually in an unsatisfactory value-engineering exercise that might deliver 50 per cent of the planned savings at most.

Sometimes circumstances or time do not allow much more than this, but where it is possible then the DMer should consider making a VM strategy and process part of the overall project process right from the start.

The culture and platform this provides will enable more satisfying project outcomes, benefits for the customer and a starting point for innovation.

The Kano Model – 'The Delighter'

At one of our workshops Michael Graham introduced us to 'The Delighter' – the Kano Model, which offers a framework for thinking about exceptional customer value. His notes are included below:

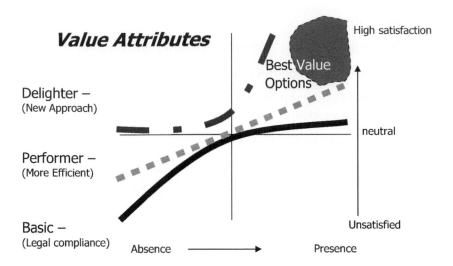

Figure 6.2 The Kano Model. Source: Presentation by Michael Graham. Reproduced by permission of Michael Graham, UKValueManagement.

About the Kano Model

The 'Kano model' suggests that value attributes of a product or service fall into three categories:

- **Basic**, essential attributes – e.g. construct to specification

- **Performance**, more of is better – e.g. maximise defect-free construction

- **Delighter**, the unexpected feature – e.g. strong synergies through construction partnering.

And the same can be said of a service:

- **Basic** – a design that complies with the law

- **Performance** – a design that is easier and more economical to build

- **Delighter** – a design that introduces unique and unexpected synergies for the client.

This 'Kano model' of what drives customer value can illustrate a complex relationship between the various parameters, but in the end the choice is often clear – there may be one distinctive (high-quality) product that meets the customer value criteria best at an affordable price, or there may be several similar products, any one of which could satisfy requirements at an affordable price – in which case lowest cost normally wins.

The strategic choice is often between maximising return from investment or minimising size of investment to achieve adequate performance. A choice made for one project in isolation may be different from the choice made for a programme of investment where synergy can be achieved between projects for minimal additional investment (frameworks). For example, property maintenance can be very expensive owing to the plethora of different lowest-cost-to-construct solutions, but difficult to maintain in installations that have been built over the years.

Whole-life assessments should consider any individual design proposal in the context of the mix of property. As another example, consider private-sector housing – a small additional cost secures a great return in terms of market price for an upgrade from a three-bed to a four-bed property, but market demand is for a mix of homes, so development value is not maximised through best-value choice for each individual home.

Spending effort and money to strengthen a basic attribute does not give much return in terms of customer satisfaction. If the business wishes to improve value, it needs to focus on delivering more of what the customer wants (performance) or introducing a new feature, which delights the customer. And for the business to be more profitable, it needs to do this while achieving better value for money in the supply chain.

– Michael Graham

7 People

The CIOB Design Manager's Handbook

'Two things are infinite: the universe and human stupidity; and I'm not sure about the universe.'

– Albert Einstein

''The most basic of all human needs is the need to understand and be understood.'

– Ralph Nichols

''Never underestimate the sheer stupidity of people.'

– Anon.

''It has been said that you can either please all of the people some of the time, or some of the people all of the time. In my experience even pleasing some of the people some of the time is an achievement!'

– Anon.

Introduction

The rate of change in our world is increasing day by day. We are overloaded with information, and every day there are new 'must-have' gadgets and gizmos. If Facebook were a nation, it would be the third largest in the world, behind only China and India, with a current population (active users) of over **900 million**, which just shows how the technology of social networking is impacting on us all and how fast life is changing. Facebook was founded only in **2004**.

An ancient Greek would take some time and difficulty to get up to speed with the technology that we now use every day without even thinking about it, but she would easily and immediately recognise the same modes of human behaviour: ineffective communication, anger, frustration, rage, jealousy, personal and work politics and the whole range of human behaviours that we all know so well. Some things, or to be more exact, people, do not change!

Successful projects are the result of teams working successfully together. Buildings are commissioned, conceived and built by people and are not just the result of applied technology. Our industry is all about teamwork and how people come together and work together as a team to achieve a common goal.

It is true that the processes, tools and technology employed will all be factors in achieving the end result, but even with the much heralded industry panacea of BIM on the horizon, these are still only tools, used by people after all. And it is this group of people, the project team, who will ultimately determine the success or failure of your project. Ignore this basic and simple fact at your peril!

The Design Manager's Handbook, First Edition. John Eynon.
© 2013 The Chartered Institute of Building. Published 2013 by Blackwell Publishing Ltd.

Admittedly the project-team leader and their particular style of leadership are hugely influential on team success, but there is a lot of material already out there about this subject. I suggest you read some books by Jack Welch, Jim Collins, Seth Godin and Simon Sinek, to name just a few.

Our human-ness affects everything we do, (of course!), but most of the time we're not thoughtfully conscious of this in the present moment, the now. You can't plan for stupidity and the vagaries of human nature, but you can take a few steps to arm yourself for the fray. While I would not pretend to be an expert in 'soft' skills, and I've had more than my fair share of failures, getting some understanding of the dynamics of teams, how they work and some of the things that make people tick can only be of help in your daily DM role.

This is especially useful when it comes to negotiating a deadline with someone who does not particularly want to help, or you're on the receiving end of some unpleasant stuff and you're just trying to work out what's going on.

So I want us to take a look at this in a few layers, rather like peeling an onion!

Firstly, we take a look at some global-level factors. We will look at some forces for change – or tsunamis, as I have called them. We'll also look at them in more detail in Section 11: Future.

Then I want to take a look at the tribal nature of our industry and how it affects the way we work together, since it determines some of the frameworks within which we operate and the corresponding roles that we take.

That will be followed by the business organisation and project-level team dynamics.

And then there's yourself, the person, and the different hats you will need to wear. Sometimes we need to be particularly adept at changing hats very quickly!

With the help of Saima Butt, we will look at some simple personal dynamics, which are expanded in Saima's paper 'Style behaviours and leadership moments' at Appendix F.

Then, to close this section, we will draw things together with a few thoughts.

Changing times

As I have alluded to elsewhere already, we are living in challenging times, and never more so than at the moment. The construction industry is currently trying to emerge from its deepest recession for many years – something unknown to the younger generation – but, sadly, the veterans have seen it all before. It is only a question of how much longer and how much deeper it's going to get. The recovery, when the 'green shoots' finally emerge, will be a very long haul. It will be several years before the industry reaches pre-recession activity levels and by that time our industry will be very different.

Technology, and BIM in particular, will have a big say in the future of the industry, as will climate change, generational change, the depletion of natural resources, training and education challenges and the skills gap, changing practices in procurement and so on.

These drivers and more are the external forces, or tsunamis, which sometimes are the push behind the agendas and the people we meet every day. They are hitting us now and will continue to impact over the next decade and their effects will trickle down to the grass roots of our industry, changing the way we work, the

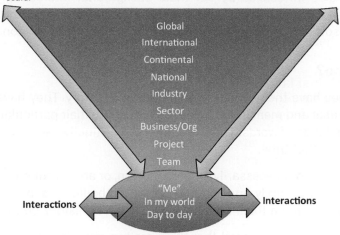

Factors, issues and influences filter down and impact on us everyday from every level – affecting our interactions, and the stage on which the DMer operates.

We can also push back up, affecting trends at every level, even on the global scale.

Global
International
Continental
National
Industry
Sector
Business/Org
Project
Team

"Me"
In my world
Day to day

Interactions   Interactions

Figure 7.1 Factors triangle.

resources we use, the buildings we construct and the environment we leave behind for the next generation. We will discuss these in more detail in Section 11.

These external forces filter down into our daily situations – sometimes determining the positions or stances that we, others or our businesses, take up in situations and meetings, or influencing the decisions that are made. A simple example is when the fees or the profits have run out, meaning that whatever happens, a particular organisation cannot commit any more resources to a task, whatever you do.

Many factors drive people's agendas and things are rarely what they seem on the surface.

Tribes

Let's think a little about our tribes.

'May you live in interesting times,' as the Chinese say. Our industry potentially will change more in the next 10 years than in the last 100, through the impact of technology both on and off site. The long-term effects will lead to more integration and collaboration, a convergence and blurring of some roles, and potentially the end of some disciplines and the creation of new ones. Could this be an example of Darwinian evolution in action? Certainly the way we have worked in the past will no longer serve us in the truly BIM-data-rich collaborative environments.

So can designers and constructors really integrate? Or to put it another way . . . if constructors are from Mars, are designers from Venus? I think they can work together successfully, but it is rather like porcupines making love – it has to be executed very carefully!

The industry in which we work is traditionally adversarial. Even with partnering, frameworks and supposedly integrated processes, we can still easily revert to type and retreat into our corners, either when feeling threatened or when given the opportunity.

We can be more concerned about protecting our own positions rather than working together – sometimes to the detriment of the projects we're delivering, and of our customers. These silos or positions can be based on contractual boundaries, but are also influenced by our tribal background. Interestingly, even within teams these sorts of dynamic can undermine collaboration and make the team ineffective – a sort of implosion, eaten away from the inside by personal agendas.

What makes a tribe?

Tribes have their own values, culture and history. They have their own particular mindset and mentalities. These can extend to their particular use of language, and even kinds of words. There is something distinctive about a tribe, and about being part of that tribe.

A tribe is not necessarily an organisation, or an institution – perhaps more a cross between a family and a movement. It embodies a sense of belonging – both a personal and a tribal identity. It is a fusion of history, culture, education and life.

So I would like to posit that there are two main tribes in our industry, with some subsets, and they contribute to some of the basic drivers that determine how we work together (or not!) – Design and Construct.

If you want to think about the sub-tribes and their mentalities, then consider:

- Do architects think they're better because they have the ideas?

- Do QSs think they're better because they control the money?

- Do engineers think they're superior because they make it stand up and buildable?

- Are bridge engineers quietly smug because they design massive complex structures?

- Are builders really the best because they turn all those drawings into actual physical reality?

- Are the specialist subcontractors the real kings of the process, because they provide the real expertise to make it all work?

And so on . . .

Each tribe has its own culture, view on life and perceptions of the other tribes (professions, disciplines, trades, etc.).

So let's introduce the two main tribes.

The tribe of design

The hallmarks of this tribe are creativity, visual awareness and intuition, and they are iterative in terms of process. Designers are prone to be more ambiguous, thinking more about possibilities, options and ideas rather than dealing in certainties. After all, they're the creative types! They view a line on a drawing as something that could change: it has potential, possibilities.

Contrast this tribe with . . .

The tribe of construct

This tribe thinks in linear, practical, cost-driven ways. They are much more concerned with deadlines, certainties, facts and the bottom line.

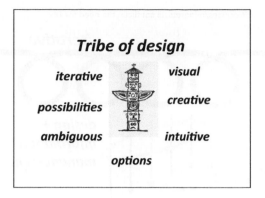

Figure 7.2

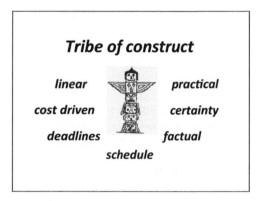

Figure 7.3

Constructors deal in certainty. After all, they have to build the project to an agreed time and cost. These are the pragmatic, practical people who work to a programme, and build with relentless logic, steadily, step by step. They view a line on a drawing as something fixed, known and to be built.

Processes in conflict

As discussed above, the process for the tribe of design could be illustrated by a loosely coiled spring, an iterative loopy design process hopefully going in the right direction – but sometimes it doesn't. The design process can get sidetracked, disappearing up cul-de-sacs, only to go forward again before closing in on the final solution.

In contrast, the tribe of construct uses linear, logical, practical process, step by step methodically assembling the building.

It is no understatement to say that these processes are fundamentally in conflict, effectively at war with each other. In addition, designers design holistically; they design the project as a whole, everything considered together. In contrast, most projects are procured and built through subcontract packages, dividing the project down into the component or element specialist subcontracts.

In terms of these two main tribes we could even be talking about different kinds of people. Naturally people are drawn to different aspects and different roles in the industry. These factors only serve to add to the divide and the lack of mutual comprehension.

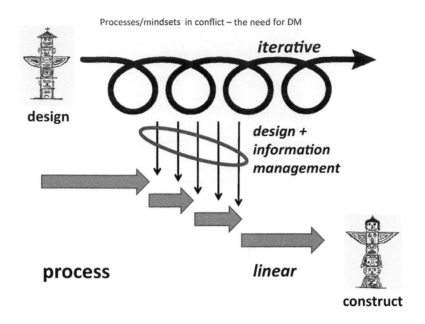

Figure 7.4

That is why and where Design Management becomes relevant to bridging this divide. To reinterpret something earlier, if designers speak Venusian, and contractors speak Martian, then Design Managers speak both!

The role of the DMer, the processes and tools they use, ensure that the design iterations produce the right information at the right time to feed the linear, practical construction process, which must not be delayed and must deliver on time, on budget, to the required quality.

However, here we are as the project DMer sitting on the uncomfortable fence between the two tribes and the two processes, which are in fundamental conflict. Even the thought processes are totally different. This divide is the result of years of training, ingrained cultural understandings, prejudices and predilections that affect the way we work, make decisions, and interact with each other in the team.

Transition

However, the tide has turned, and is now running up the beach. Some aren't aware of the immensity of the tide of change that will sweep across the industry, leaving virtually nothing untouched. The recession has brought things into sharp focus: the drive for efficiency, cost reduction, technology, zero carbon, climate change, resources, demography, skills, training – all are slowly tightening their grip.

Our industry is under huge pressures to change. Under the pressure of this onslaught, I think the tribes are in transition. They have been for a while, and the process will begin to accelerate exponentially. There will be vested interests that try to maintain the status quo of the current positions for their own reasons, but I suspect the force for change will in time become irresistible, affecting disciplines, education and eventually institutions.

Undoubtedly there is a lot of navel gazing going on in the industry at the moment and also in the various disciplines. Roles and training are under scrutiny – particularly at a time when the costs of education are soaring and graduate unemployment is on the increase. The relevance of graduate training to businesses and industry is being seriously questioned. Students are asking whether they are obtaining value for money in the training they receive, particularly as they will be in debt for most of their working lives to pay for it. Practitioners question the quality of graduates being produced and whether they are fit to practise as professionals.

At the same time, the industry is under greater pressure than ever to reduce costs, invest and innovate, maintain competitive edge and simply stay in business.

These external, global, societal factors, plus the tribal background, provide a backdrop to the project environment that the DM operates within, the roles that people play and the positions that people can take up in conflict, negotiating teamwork and leadership positions.

One tribe

I just wonder whether the time has come for us all in the industry to think of ourselves as one tribe. I couldn't come up with a mind-blowing name, but . . .

The tribe of solutions

Whether architect, engineer, contractor or subcontractor, we all contribute to making and delivering solutions for our customers. We need each other. We need to work together and ultimately, upon every project, our fates are inextricably intertwined. We succeed or fail together.

We need to think of ourselves as one tribe, and to value the distinctive contribution of each.

However, there is a major problem, and for every one of us that is in our DNA. It takes years of training to eradicate the gene that causes these problems, and even then the cure may not be successful!

It's in the DNA

This process begins at our birth – that is, our professional 'birth'.

Our tribal culture starts in the colleges and universities where we complete our qualifications and degrees, and on the sites where we train future generations.

Isn't it ironic that we train our future professionals in tribal silos, reinforcing the generations of history and tribal culture? Upon completion of their courses, graduates are released into the world 'to go forth and integrate'! – work together, one team, with common goals and understanding. It is as if this will happen suddenly and work by some mysterious process.

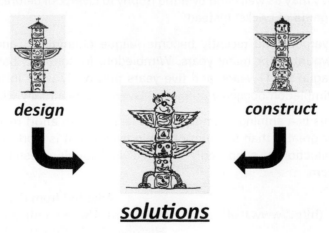

design   *construct*

<u>solutions</u>

Figure 7.5

7 People

117

As we have seen, tribal understandings can act as filters on our interaction, and we haven't even considered basic personality issues yet! Even when, in recent examples of industry best practice, teams have integrated, collaborated and communicated successfully, they can then still revert to adversarial type on their next project with no apparent difficulty.

Clearly something does need to change. There are many reasons why, but I think these forces for change, both external and internal to the industry, are conspiring and converging and within a generation will force us to adopt new ways of working, to change education systems and perhaps even to change institutional structures. BIM is a huge opportunity for the industry to move forward and is one of the main drivers, combined with the economic drivers of society.

Within the disciplines, this isn't just about architects. I attended a conference for specialist engineers a few months ago and I was amazed to learn that a lot of the issues discussed were just the same as the discussions raging in architectural circles, such as their position and standing in the industry, relationships between designers and constructors, the skills gap, relevance of training to practice and graduate unemployment. Similarly, from postings on the internet recently it can be seen that surveyors, too, are undergoing some heart searching about their role, their future in the industry and their institutions and groupings.

A tribal understanding of our industry and these external forces provides a few layers or filters on our interactions. The dynamics of businesses, our teams and our own personalities and preferences are additional factors in affecting how we work together.

Teamthink

It has always fascinated me how some project teams composed of talented individuals can be outperformed by a team of supposedly less able people who work together better as a team and achieve outstanding success.

> To use a sporting analogy, Wimbledon Football Club in 1988 won the FA Cup at the expense of the supposedly superior Liverpool – a classic example of a collective team performance winning over against individual talent.
>
> Before the game they were 33–1 outsiders to win. I remember Bobby Gould their manager at the time saying that if the pundits were to be believed, then they may as well hand over the trophy to Liverpool before the game, and have a game of cricket instead!
>
> Liverpool had recently become league champions, and were the hottest favourites for many years. Wimbledon, in contrast, had only been in the league for 11 years, and five years previously were in the Fourth Division. Wimbledon doggedly outplayed Liverpool, eventually winning 1–0.
>
> This is a shining example of a collective team effort resulting in something far greater than the sum of the individuals. It is worth remembering that a collection of supposedly 'talented' individuals doesn't necessarily make a team.
>
> Adapted from *Classic Cup Finals: 1988*
> (http://www.thefa.com/Competitions/FACompetitions/TheFACup/History/
> HistoryOfTheFACup/1988WimbledonLiverpool)

Leadership would be a factor here, and certainly I've worked with great team leaders in the past. Leaders who know how to get the best out of the team bring people together to a common purpose, and also sometimes have the ability to say just the right words for the right moment, leading the team to achieve outstanding results in the process. On one project there was a project manager who without doubt was the best I've ever worked with – the project was a resounding success and also very profitable, and to a man I think we would have walked through fire if he'd asked us to!

However, it is the composition of teams that also isn't given much deliberate thought, other than whoever happens to be available for a particular role at a particular time. Perhaps more thought could be given to how we put teams together, not only balancing skills, experience and expertise but also balancing the fit of people, to give the team the best possible start and the collective tools for the job ahead. Where that isn't possible, a little more awareness around the subject will help teams to collectively work together through a mutual understanding of where problems could arise.

Dynamics

I would like to touch upon psychometrics and personality type. There are many profiling model systems out there, probably most notably Myers Briggs, Belbin and KAI, to name just a few. We have included a few ideas for further reading in the bibliography, but have tried to avoid using the terminology of any particular system and have taken a more neutral approach.

However, a few words of warning. I am sure you will agree that people are pretty complicated! Sometimes what you see is not what you necessarily get, and the danger with modelling behaviours is that it can quickly descend into stereotyping. People do not fit into boxes as much we like to think they do, nor do they behave according to the labels we assign to them. I prefer to think of these systems as indicators of behaviour, but they are no more than that. Nothing is rigid. Be ready to be surprised, as the range of human behaviour is far wider than we can sometimes imagine!

Personality types, or indicators of behaviour, are just naming the preferences in the way people *prefer* to behave. This doesn't rule out their behaving outside their 'type' in certain situations – because that's what people do!

Also be aware that as people grow and change through their lives, the subtleties of their profile can change also. Major life events can have a significant impact upon our preferences – for example, bereavement, marriage, divorce, redundancy and so on. I know from my own experience of having completed Myers Briggs profiles about 12 years apart. Some of my preferences moved across boundaries into different categories. In those 12 years I had experienced some significant changes both at work and at home – they had an effect.

There is another aspect in team and project life, which I want to mention in passing.

Fun and joy

(*Enjoyment, amusement or light-hearted pleasure. A feeling of great pleasure or happiness* – Oxford Dictionary of English)

As much as our human-ness affects what we do and how we do it, so I think we need to be deriving some pleasure – joy, fun if you will – in the process and the

7 People

outcome. The best teams for all the ups and downs of working relationships actually *enjoy* working together. Yes, there'll be some banter, but underneath there will be mutual respect, openness and integrity.

The cogs will be eased by some social sessions, whether this is at the pub, over meals out or while paintballing. Just a thought, but has it all suddenly become very serious? Of course the stakes are high on any project, but it all becomes just a little easier if we can enjoy working together as well.

> As Paul Phillips put it in a recent discussion on the Design Managers Forum: 'Make money, build safely, delight your client, save the planet and have fun on the way.'

Know yourself

If this subject of people is of interest to you, it is worth taking a few profile tests to get to know your own indicators and also to get to know a few of the systems. It is interesting to see whether the results chime with your own perception of yourself, or provide a few surprises. With a little knowledge it becomes relatively easy to profile individuals in a rule-of-thumb fashion to analyse simple dynamics between people. Occasionally this has helped me to understand others and to help with communication and team working.

This brings us back to the role of the DMer. Consider the classic exchange between an introvert and an extrovert, or say a creative type with a more logical practical type. What is happening? Different things are important to different types of people; the processing of ideas and thoughts is happening differently. For the extrovert it happens more externally, while the introvert will tend to say much less, but the eventual statement will be more considered and concise.

Add to this the different hats you will need to wear, sometimes the diplomat, sometimes conciliator, sometimes the negotiator and so on.

So how well do you know yourself and what is going on around you?

Values

Who are you? What do you stand for? The DMer stands in that uncomfortable zone between designer and constructor – and many other stakeholders, for that matter. A major part of the role is building successful relationships that will last the lifetime of the project and perhaps beyond. This is critical to your success. Values like respect, trust, integrity, transparency and honesty spring to my mind. This doesn't mean agreeing with everyone, or not saying no, nor giving away sensitive information or the crown jewels, but it does mean being the same person to everyone, and performing your role with integrity.

Over a project's duration there will be give and take, ebb and flow in all relationships. Sometimes you will need help; sometimes other members of the team will need your help. The best projects work without having to refer to the letter of the contract. This is where the whole team comes together and collaborates, crossing the boundaries of their respective businesses, and their roles, to achieve the project goals – effective relationships in action. Everyone benefits from a successful project – likewise everyone shares the pain of failure in some way.

So what are your values? What is important to you? If you know yourself, your ability to be authentic and to build lasting and productive relationships will increase, benefiting yourself, your team and your project and adding value to your business.

And finally for this section

While process, tools, and checklists are the cogs of DM, we sometimes forget that the people using them are not robots, nor are they machines. This human-ness factor colours all our interactions and provides many layers of meaning, agendas and interpretations. It affects everything we do, how we communicate, how we work together – or don't, in the final analysis. This aspect is so important and yet so frequently overlooked. It is the oil that keeps the cogs turning efficiently and smoothly.

I am hoping that we have lifted the lid a little on the 'softer' side of DM!

To be successful in DM, or any role for that matter, we need to learn to understand how people interact and relate. Understanding ourselves and what we bring to the project, with our own distinctive traits and giftings, is also a key factor. Saima explores this in more detail in the related Appendix.

8

Training

The CIOB Design Manager's Handbook

'Until you have worked for a main contractor, you will not have the breadth of experience to fully understand the nature of the task . . . it really does help if you have an understanding of the construction sector as a whole – contractors, clients and their stakeholders and the social/economic/political framework.'

– Phillip Yorke

Introduction

And when I grow up I want to be a Design Manager . . .

When Design Management was starting to become prevalent in the 1990s, then you could find all sorts of people doing the role. You still can now, of course, but back then there weren't the courses that are currently available.

Where do Design Managers come from? How do you become a DMer? What qualifications do you need, how do you train?

People enter this role from all sorts of backgrounds. A few examples:

- A site manager is asked to 'look after the drawings' and begins the journey of managing information, getting involved in supply-chain issues and eventually becomes a site-based Design Manager.

- An architectural technologist working for the architects on a project so impresses the main contractor that he is invited to join the contractor's team and co-ordinate and manage design production.

- A project manager decides he wants to take a different direction and, using all his skills and experience in managing construction projects, becomes one of the best delivery Design Managers I've ever known.

These are real people. And there are many more – engineers, architects, surveyors, admin staff have all found a route into DM.

In the past perhaps DM was seen as something that needed to be done; it was thought that anybody could do it, and was just dumped on the nearest available pair of hands.

It is then a short jump to becoming the fount of all knowledge on the design information on the project. 'Do you know which drawing has the details of the widget cupboard and the ironmongery schedule to go with it?' The number of times I have found the drawing myself, simply because by the time I've explained

The Design Manager's Handbook, First Edition. John Eynon.
© 2013 The Chartered Institute of Building. Published 2013 by Blackwell Publishing Ltd.

where it is, and how they could find it for themselves, and got them to do it, I would have had time to review ten subcontractor's drawings!

So this idea of there being someone to draw the design threads together isn't new; it is perhaps that in a way we have fallen into it. I imagine in the early days there was the assumption (hope?) that the designers would just understand and produce what contractors needed and wanted, as if by magic. This didn't happen – rarely, anyway. That was perhaps as a result of two things – firstly, designers not really understanding what contractors and supply chain needed at different stages of procurement and construction; and secondly, contractors not being able to articulate exactly what they wanted and needed and why. So in the end the contracting side had to man up on design, perhaps reluctantly, but it was necessary to in order to get the required information on time, deliver projects successfully and protect profit margins.

Roll the clock forward and DM is much more accepted now across the board and is expected to play a part in project success. I suspect that even some designers reluctantly accept that having someone there who is helping to get their information packaged and flowing in the right way can't be a bad thing, even it means some devolution of responsibility.

I am speculating, but probably the majority of current DMers have come though the contracting ranks somehow, either off the tools, site-management roles or something related. Engineers also make up a large cohort, and I think this reflects that engineers generally have always stayed a little closer to the construction process than architects. Is it coincidental that as the architectural profession has moved to an increasingly academic style of education and training, their influence over the design/construction process has waned?

But now, as DM courses have developed over the last few years, we have graduate Design Managers coming to the coalface, with current academic thinking and a willingness to embrace new ideas and technologies.

That has made me think: is Design Management really a career choice? Should it be? Is it possible to manage a process without fully understanding all of the technicalities and mechanics? It takes several years for any building professional to become proficient after graduation, and a graduate DMer will be no different.

A great DMer is expert in some things, but a jack-of-all-trades in everything, understanding how it all comes together. However, current graduates are going to have another hill to climb as the tectonic plates of our industry shift owing to the forces of economics, and the cultural and technology drivers of BIM.

Today's DMers, tomorrow's BIM managers? Perhaps – but it is going to be a steep learning curve for us all, and DM could be just right in the middle of it.

A role to aspire to, after all!

Training

So how do people get *trained* as a DMer? Graduates may be *educated*, but *learning to do the role* is something else.

An accepted pattern seems to be starting at the delivery end and working backwards. So a trainee Design Manager will be teamed with an experienced DMer on site and will go through the process of completing the construction phase.

Having perhaps gone through this stage a few times, they will then be on their own project, perhaps again a few times.

At this point, it then depends – the developing DMer can either continue in site-delivery mode or move back along the process chain into pre-construction, working on pre-contract design, tenders and the like.

Depending on your background, each stage of this process will present different challenges. If you are a designer, then you probably won't have experienced the intimate workings of a site team on a daily basis, looking at safety, logistics, programming and procurement blow by blow, for example. Equally, if you're from a trade or site background you may not have much experience of how designers work, iterations, reviews, stakeholder engagement, statutory compliance, etc.

It takes time to fill in these different areas of experience in order to produce someone who has a rounded knowledge of the entire process – not necessarily in exhaustive detail, but enough to understand each stage of the process, the people involved, what needs to happen, etc.

Paula Bleanch of Northumbria University has been looking at this subject with the CIOB DM working group, examining the pitfalls of current DM education and training and how the DMer of the future will be trained. Her paper 'Educating the Design Manager of the Future' is included as Appendix F.

And finally for this section . . .

Bringing focus to Design Management is rather like shepherding cats, to use a northern phrase – quite difficult.

Let us consider a few factors:

- There exist a variety of pathways into the role.

- There is a wide variation in understanding of the role and requirements, together with job titles and descriptions.

- Practitioners have a huge diversity of background and qualifications.

- Application of DM process can vary in practice and terminology.

If nothing were to change in the industry in the short term, then bringing this together should be readily achievable within a short timescale. However, as I discuss in Section 11, the rate of change is accelerating and our industry is undergoing probably the greatest transformation since the Industrial Revolution.

So DM and the DMer are in transition, and with the impact of BIM, that will only increase the challenges for our educators and trainers.

9

Quality
The CIOB Design Manager's Handbook

'The standard of something as measured against other things of a similar kind; the degree of excellence of something.'

– Oxford Dictionary of English

Introduction

As the project DMer, you will meet up with 'Quality' in various guises in the life of a project and you will need to respond appropriately.

We can consider the quality of the project process, and the quality of the delivered end product. We can talk about the quality of the project experience for all the stakeholders. We can also look at the quality of what we have provided in terms of the visual and design quality, the quality of the build, how it is constructed and its impact on the environment and the urban fabric, if that's relevant.

Following on from this latter point is the question of the quality of the legacy that we're leaving for following generations, and I wonder whether it is (or will be in the future?) enough just to say that we did what we were contracted to do, or that we were simply doing what we were instructed to do.

In terms of our own integrity, is 'just enough', or 'compliant', actually good enough?

Perhaps another aspect to quality is to consider sustainability (social, economic and environmental) in terms of successive generations.

Let's take a look at the simpler aspects before returning to legacy issues for the DMer.

Process

Looking at process takes us back to the start of our journey. Before we consider the How, we need to be thinking about the Why and the What.

A principle of working to ISO 9001 for a Quality Management System is to set out in your process manual exactly what you do, and then to do exactly what you have said you would – that is how you will be audited.

As we have discussed in Value Management, we need to understand our customer in depth, to be able to make sure in following through the process that we are meeting their drivers and objectives. The degree to which we meet these factors will determine how successful we are in achieving the required quality.

The Design Manager's Handbook, First Edition. John Eynon.
© 2013 The Chartered Institute of Building. Published 2013 by Blackwell Publishing Ltd.

This brings together alignment, engagement and communication. We need to be in that place of understanding all of these factors and the whole team needs to be aligned on the objectives, the desired outcomes and the underlying drivers that are determining our course of action.

This is a key role for the pre-construction DMer, through workshops and discussions, ensuring collaboration and understanding at every level. Remember that at this point some of the team are speaking Greek and others Latin (!), so it is worth spending some time to make sure that there is common understanding across the whole team.

Design Quality Indicator

The DQI developed by the CIC arises out of a demonstration project from Sir John Egan's Movement for Innovation (M4I): www.dqi.org.uk.

Unfortunately, as a tool it has not been used as widely as it could have been. However, it does provide a useful way of establishing the right conversations and level of understanding at the various stages of a project. Through workshops and questionnaires which are facilitated, the DQI can be used to develop the Brief, review the Design as it develops, review the built product and also evaluate post-occupancy issues for feedback.

The beauty of the DQI is that through using the process, you can establish a common language for the project right from the start, and customers, designers, end users, members of the public, can all be discussing drivers and requirements for the project with mutual understanding.

The DQI looks at this under three main headings:

- **Functionality**

- **Build Quality**

- **Impact.**

Functionality

- is concerned with the arrangement, quality and inter-relationship of space, and the way in which the building is designed to be useful.

Build Quality

- relates to the engineering performance of a building, which includes structural stability and the integration and robustness of the systems, finishes and fittings.

Impact

- refers to the building's ability to create a sense of place, and to have a positive effect on the local community and environment. It also encompasses the wider effect the design may have on the arts of building and architecture.

(source: DQI guide – CIC)

The tools are tailored to the different stages of the project, and stakeholder responses are processed using the online tool to produce output displays. This is explained in much more detail on the website.

Once the briefing stage has been completed, then subsequent stages are assessed against the brief outputs, to highlight variance and inconsistencies. These can then be explored in more detail with the team.

For me, facilitating sessions has been very interesting, simply because it is where the differences in opinion or scoring occur that are worth exploring. As discussions develop, they then enable the requirements and expectations to be exposed to the daylight, explored, opened out and then pinned down.

For more on the DQI process, finding a facilitator, and the tools, just go to the website: www.dqi.org.uk.

Post-Occupancy Evaluation

An area that has been sadly lacking is the POE. Teams are very good at walking away on completion and, while for them there is the next project to look forward to, for the building owner and operator life will go on for some considerable time!

All that good work in design and construction, in building relationships, working with stakeholders and aligning expectations, can be undermined, undone and wasted, as post-completion is where it can all begin to go wrong.

Looking at this from a change-management perspective, moving back into a refurbished or new building, with new systems, room arrangements and facilities, can be both a daunting and exciting prospect for the building users. Inevitably there will be teething problems, but with the right levels of support and pro-active defects management, it can be effectively carried out.

BSRIA has carried out a lot of work in this field, and the 'Soft Landings' process provides a route map to making completion, handover and operations as smooth as possible. Included within this is Building Performance Evaluation (also known as POE), which considers functional performance, energy consumption review, and occupant satisfaction levels. The use of DQI or the BSRIA process can provide constructive feedback useful to teams in delivering their next projects. See

www.bsria.co.uk/services/design/soft-landings/
and
www.bsria.co.uk/services/fm/post-occupancy-evaluation/

For all its ills, PFI procurement in the UK has brought project teams much closer to the operational stage of the asset lifecycle and has brought proper consideration of the operational life of an asset into the value equation.

Most contractors have recognised that the handover, aftercare and operational stages are just as important as other stages of the project cycle, and have specialist teams and processes for this stage. A happy customer and end users is good for business – a lesson we all could learn!

The quality envelope

Referring back to the Four Stage Process (see Section 2), we have defined 'the project envelope' at the point of formation of the delivery contract, where pre-construction project procurement ends and project delivery starts.

This sets up the delivery phase parameters or boundaries and consists of:

- Time – the agreed programme
- Cost – the agreed cost
- Quality, consisting of Function, Content, and Standard.

Function

The design should address the use of the asset, relating back to the defined brief. What is it the customer wants the asset to do? How will it be used? And so on, leading us to . . .

Content

. . . which consists of the facilities, equipment and systems that the asset is to contain: both base build and FFE if required.

Standard

Gold-plated or galvanised? Sound insulation 50dBA or none? Air tightness $5\,m^3$ or $10\,m^3$? – establishing the standards of the asset in terms of technical criteria, or standard of component or finish. Fair-faced blockwork, or blockwork finished fair? (There is a difference!)

The standard will be established both by the design drawing and by the specifications and supplementary information.

The overall quality as described by the design information should be back to back with the defined brief. By the time you have reached this point, the scheme design should been reviewed against the brief and agreed.

It is far easier to change the design than the building, so making sure the expectations and proposed delivered product are in alignment will save problems later, such as last-minute changes during manufacture and construction and unfulfilled expectations for the customer.

The better the definition of the delivery envelope at this point, the more straightforward the delivery stage will be, and the objective of achieving expectations (which should have been aligned by this point) will not be an issue.

As the delivery phase progresses, you will soon get a feel for how good a job has been done on this – the content of ongoing meetings, workshops, conversations, correspondence and emails will give you some clues!

In Section 4: Tools we have some considered some techniques that are helpful in defining quality – visiting completed projects, reference projects as benchmarks, samples, prototypes, mock-ups, visuals, animations are all useful ways of getting people to see and understand what is on offer. A picture paints a thousand words, and for an industry that is so concerned with the visual and experiential, it is surprising how little we use this aspect to clarify requirements.

'Go and see, touch and feel' – understanding becomes quicker, more accurate and far less ambiguous.

Design and build

From what I can see on the LinkedIn Design Managers' Forum and from talking to people across the industry, I think it is probable that the majority of 'Design Managers' work for contractors and client organisations. While there will be those in design practices, the DM aspect there will usually form part of another role, say job architect or project leader.

How did we get here? It seems that as design-and-build became more prevalent as a procurement route, then in turn it became clear that someone was needed to help manage the flow of design information through to the contractor and their supply chain – hence the rise of the Design Manager and Co-ordinator. Over the last decade we have seen the prevalence of contractor-led procurement, principally because it brings the delivery team much closer to the designers, and increases certainty around the delivery phase, while hopefully reducing project risk for customers.

The trend towards more integrated models of delivery will continue, whether contractor-, PM- or designer-led. However, it has brought contractors into a realm where traditionally, in the past, they were more on the receiving end rather than actively participating and leading the process – namely, the design of buildings and assets, and their impact.

PFI, as we discussed earlier, has brought into play more consideration of the life cycle, economies and value.

Design integrity

D+B has had many detractors over the years – particularly with designers complaining over cheapening of their designs, or the dumbing-down of quality. Certainly this may have been the case in the past, but I don't think the argument holds so much today, as we have discussed.

There are many examples of high-quality buildings delivered through D+B styles of procurement, with leading designers maintaining design integrity and quality and winning awards in the process. The DMer can have a positive influence on the design process, assisting the designers in steering through the requirements of the contractor whilst maintaining the core concept and ideas. It's not all negative!

Failure of D+B, when it occurs, can be traced back more to how the project was set up and defined. If the Employer's Requirements have gaps and inconsistencies, then unfortunately these will tend to be exploited. Contractors are trying to win work under the rules of the project set by the customer and their team. Definition at the outset is the key, and if this is found wanting, then issues will arise later.

As we have also seen, pre-contract cost planning is critical. Sometimes teams have an inflated expectation of what can be achieved, because the pre-tender cost plan is not in line with the current design. The tender-winning contractor is the bearer of bad news and painful value engineering can be the result. it's not a good way to start the delivery phase!

Impact and legacy

Most major contractors are now exemplary in their approach on Corporate Social Responsibility, sustainability, the environment and health and safety. Indeed, they have to be, as these are frequently part of the scoring criteria for awarding work. So in terms of the delivery process, it is broadly recognised that contractors have much to contribute – how they impact the local community, provide employment and training opportunities, and also minimise harm to the environment, making the whole process better.

Surely the challenge now is to go much further? Climate change and carbon emissions, energy and resources shortages, demographics and a host of other factors will influence and drive the types and kinds of building that we design and build over the next 50 years. How do we make the end result better?

We are all part of the problem, but also all part of the solution. Designers, constructors, supply chain and all the stakeholders: together we will make the difference – the Tribe of Solutions.

> *If over 70 per cent of the buildings that we are constructing now, and those already existing, will still be standing in 2050 and beyond, what sort of legacy are we leaving to future generations?*

- What impact are the homes, workplaces, civic buildings that we produce having on people and their daily lives?

- Are we making life better?

- Do we care?

It is these sorts of issues that intelligent contractors should be able to discuss equally, with architects, planners and urban designers.

Certainly as contractors have skilled up and developed or brought in in-house design and design management expertise, the quality of dialogue has risen. The broadening remit of corporate responsibility and sustainability has brought the contracting fraternity right into the centre stage of action on design, environment and legacy.

The professional DMer, speaking Venusian and Martian, is ideally placed to take the dialogue forward, particularly appreciating the impact that any building, infrastructure or urban project has on society at large and speaking with authority across the design construct tribes – understanding the language, the processes and the industry and the issues.

We are all playing a part in this, no matter what role we have in the project, and therefore have to take a share of the responsibility; it cannot be avoided.

And finally for this section

As integrated models of procurement become the norm, and are potentially enabled by BIM environments, then the contractor with their supply chain must become an integral part of the team from the beginning. Teams will become more cohesive, collaborative and responsive. The DMer will have a key role in this process going forward, acting as the lynchpin for project collaboration and crossing the divide between design and delivery.

Quality has many aspects, as we have seen. The DMer can be the champion of quality – whether that is maintaining adherence to project standards, protecting the 'design' or articulating the discussion around our wider impact as an industry on society at large.

---

10 Stories

The CIOB Design Manager's Handbook

And you may find yourself in another part of the world
And you may find yourself behind the wheel of a large automobile.
You may find yourself in a beautiful house, with a beautiful wife;
You may ask yourself, 'Well, how did I get here?'

– from 'Once in a Lifetime' – Talking Heads

Introduction

So . . . how did we all get here?

This section is about providing some biographies of Design Managers and also case studies showcasing some of the benefits of DM.

The WHY for this section is simple.

Firstly, it's in terms of biographies to show the diversity and background of people in DM. Whether in time this diversity will become narrower through the graduates from DM courses taking over we don't know, but I hope not. The current diversity of experience and outlook, while creating some challenges, is in reality the strength of DM.

Secondly, there are the case studies. Historically it has been difficult to quantify the benefits of DM and in some companies there is still discussion, even now, about whether DM is really necessary or provides value for money.

If you have someone responsible for integrating the design inputs on a project, making sure that the right information is being delivered at the right time to whoever needs it, as well as improving accuracy and buildability, reducing risk and adding a little innovation, then instinctively you would think this must add value. However, there will still be those asking 'Show me the money!' An answer to this is that we cannot always put numbers on this, so use your common sense!

Biographies

John Eynon

My route into DM wasn't planned – it just sort of happened! I was educated as an architect at Nottingham University in the 1970s. My training began after graduation and my original aim was to have my own practice. I moved to Croydon after graduation and passed my Part III while working at the Property Services Agency, the government design department at that time. I left the PSA and worked in a few private practices. I became an Associate with a multi-disciplinary practice and worked on a variety of projects at all stages. I also spent a year as a site architect on a large residential development in east London.

As a site architect, I had a great time working with the management contractor and subcontractors. I'd always been interested in the delivery end of the equation. Most of my time was spent making up details on site – sketches in felt pen on graph paper were my speciality. CAD was in its infancy at this stage; there were no iPads and my Apple Mac at home had a 40mB hard drive!

However, this was just as the last big recession was biting. I was made redundant a few times. The office I worked for closed. I then worked for myself and with a friend for a while, and did a few other things, including specification writing. Eventually I joined RIBA Information Services, working in information technology and research on building products.

One day I was phoned by an ex-colleague from a previous architect's practice, asking if I was interested in joining Tarmac Building as a Design Manager. I joined Tarmac, which became Carillion a few years later. I was resident DMer on a residential development and then on an office block in the City of London. It was then that I started to get involved in pre-construction and tendering.

It took probably at least 12 months or so to adjust my architect's mindset to contracting – understanding how the project teams work: not only the different roles, but also the sheer immediacy of the construction process. It stares you in the face every day! – appreciating the commercial influences, and the immediate and constant pressure of the work on site needing to progress.

I later joined Wates Group, and I had a great time with the guys there, working on all kinds of projects; now I'm working as a consultant again, on DM, BIM and process. Over this time I also helped with developing the CIOB approach on DM, BIM and carbon.

Looking back over the last 15 years or so, I think for me it has been getting the feel of and appreciating the whole design and construction process, from different sides of the table, and understanding the roles and value of every member of the team.

DM connects and brings it all together.

The construction industry is a great industry to work in and we ought to appreciate each other more.

Whatever our different roles, organisations and businesses, we all need each other in order to succeed.

Together we are all the solution.

Mal Jacobsen

I started out in the construction Industry in 1984 as Cadet Cadastral Draughtsman, setting-out building sites etc. My preference at the time was to get a cadetship in an architectural firm, but the opportunities were slim in the small town of Rockhampton, Central Queensland, Australia, where I grew up.

I then worked as a draughtsman for architects, structural engineers and construction companies in Queensland and New South Wales for the next 20 years. Concurrently, I studied by correspondence and 'after hours' in the fields of engineering and project management, finishing with a master's In Property Economics in 2010 (while always working full-time).

My drive for 'self-improvement' was due to my intention to forge a project management career in construction, rather than maintaining a technical role in design consultancy. Given my 'design' background, opportunities arose in construction companies as 'Design Manager'.

Design Management suits me perfectly – I love great design, enjoy working on construction sites and seeing buildings progress, and am extremely interested in creating and maintaining environments that support high-performing Integrated Project Teams.

Currently, I'm lead Design Manager on the $1.2 billion 'Queensland Children's Hospital' in Brisbane, Australia, where I lead a team of 20 Design Managers, Services Managers and Document Controllers.

I also lecture part-time at the Queensland University of Technology in 'Building Studies'.

Design Management is the perfect career for me, and I can't imagine doing anything else!

Alec Newing

Why design management? I have an underlying interest in things being well designed – well considered, if you like. It's in my nature to think things through. Design in this context could relate to not just products, such as built environments, but to systems and services also.

My career has developed driven by a desire to try and improve preceding activities.

While I was a site engineer and manager at the beginning of my career, I had a desire to plan the works better, to spend more time thinking operations through, before they were required to happen on site. This led me towards a combined role of planning, buying and a review of safety risks at the dawn of CDM. Consequently I became more immersed in design reviews from a construction perspective.

My first change in employer led me into a full-time planning role as a planner for fast-track retail projects. During this role it occurred to me that planning around the design process was extremely important, and indeed was lacking. I started to consider how design sub-projects should be planned and I prepared some outline best-practice guidance on the matter.

It was while working in a general project co-ordinator role on a larger design-and-build retail development that I shifted to full-time Design Management. The project revealed that the design process needed managing as a priority, as opposed to just planning the construction.

After managing design on a number of projects and bids, I wondered what was going on in design management terms in a wider context, so I secured a role as a Regional Design Manager. This gave me an opportunity to appreciate views and approaches not just from a number of Design Managers, but also from a number of other disciplines within a region.

From time to time in my career, I return to the 'coalface'. On one occasion it was to manage design for larger, more complex projects. After managing a large PFI-style project from agreed concept to six months into construction, I returned to a design management leadership role, becoming head of Design Management for a large contractor.

In this senior role, I looked to clarify design management in the organisation, rationalise the resource to have suitably qualified and experienced people. and address corporate governance for design management.

Appreciating that there is more to managing design than doing such for a main contractor organisation, my latest role is with a support-services business that procures design and construction. It is a move to better understand and support design and construction clients.

As for built-environment design management, I use my construction knowledge and understanding to inform better design solutions. I view my current role as follows, which I think is a good basis for design management generally:

- Design the process

- Facilitate decisions

- Hold designers to account

- Lead the evaluation of options

- Help designers to understand the business . . .

- . . . and the business to understand design.

Margaret Conway

I started out working for BT as a telecom engineer. I had a family, two children, and decided I wanted to go to college, as I had always had an interest in construction. I rang the course director on a Friday afternoon and had my first lecture on the following Monday morning!

I chose to take a BSc Honours course in Construction Engineering and Management at the University of Ulster, Jordanstown. I selected this course as the course director said it could lead to a diversity of construction roles.

I did my placement year out with the University of Ulster in the estates department – working as an assistant project manager client-side on some large, multimillion-pound developments for the university. This included the redevelopment of the Belfast city centre art college. This project was going through the design and tender phase when I worked on it, so I received a good grounding in the design approvals process and co-ordination between the design/services/structure fields. I was also heavily involved with working with the end users to make sure the building would be suitable for their needs. I learned a lot from working alongside the university project manager, who was an architect by background.

I returned to university to finish my final year, graduated top of my year with a first-class degree and received a CIOB prize for top student.

I then applied at a graduate fair for a job with Farrans Construction. When I went for my job interview, the director of the healthcare division, when he heard what I had been doing in my placement year, thought I would be suited to a role they needed filling on a healthcare project – there was a requirement for a site-based design coordinator on a PFI D&B laboratory project at Altnagelvin Hospital.

I worked on the project for a year, then moved on to the South Block redevelopment, phase 3.2a for another two years, again as a design coordinator. See www.farrans.com/projects/altnagelvin-hospital/

I then went on to work on the Trauma and Orthopaedic theatre development at Craigavon Hospital for 18 months: www.farrans.com/projects/craigavon-trauma-and-orthopaedic/

After that I joined McAleer and Rushe, who specialise in D&B contracts. I worked on the Aloft Hotel at Excel as design co-ordinator. The project was handed over in October 2011 and now I am currently working on PQQs and bids: www.mcaleer-rushe.co.uk/design-build-construction/hotels-leisure/aloft-london-excel/#primary

I applied to the CIOB through Dr Sarah Peace for the Tony Gage scholarship to study for an LLM in Construction Law and Practice via distance learning, in which I am happy to say I was successful. That is how I ended up on the FAS board!

In conclusion, I found the role very hard at the start, as it was a unique one and very few companies in Northern Ireland were using design co-ordinators at the time. I had to try and develop my own style, as I had no one to learn the role from.

After six years as a design coordinator I find that the most important skills needed are people-management skills and the ability to look ahead – for example, looking six months ahead in the project programme, to ensure information is always available to the construction team at the right time. Be proactive rather than reactive.

Nicholas Gill

Over the course of a 30-year career in construction and as an architect, I have experienced working in private practice on my own account, for public authorities, as a partner in a well-known firm of surveyors, and for a high-profile architectural practice specialising in healthcare buildings, before moving into Design Management.

As my career progressed, my role changed – as it does for many – gradually from designer to manager, and I found myself spending less and less time at the drawing board (or latterly, the CAD computer), and more and more time managing the activities of my staff and controlling their outputs.

This experience very quickly taught me that with increasing seniority comes responsibility, including responsibility for the content and quality of the work done by my staff. I became very adept at finding problems or mistakes in drawings and documents, and as a result developed a knack of occasionally making myself unpopular.

I realised that these skills could be more valuable to a contractor than in an architectural practice when, about ten years ago, I was working on a very large hospital project for a national contractor. I watched closely how their own Design Managers worked and interacted with the consultants and their own team. I reckoned that I could do a much better job of it than they were doing. My thoughts were brought into sharp focus, however, when I discovered how much better paid they were!

In 2006, the opportunity to move to Skanska came up. Skanska's approach seemed to me to be different from that of other contractors; with policies in place that emphasised a commitment to completing work to the highest standards, providing an integrated service employing the best designers. This was a very attractive proposition.

And so it has turned out to be. Skanska recognises and values the role of Design Manager. The firm understands that no matter how good the design team is, its activities need marshalling and co-ordinating by competent people within the contractor's team.

As a result, I've been privileged to work on projects of enormous technical complexity, with some of the best designers in the country. I believe that nowadays both design teams and contractor teams see the Design Manager as a key role in a successful project, and that the role is pivotal in keeping information flowing that is of the right content and quality, as well as to programme.

It can sometimes become a little frustrating to see younger architects doing what I used to do, and wishing I could do some more of it myself, but overall I find my role of Design Manager interesting, fulfilling, and particularly satisfying when I see large and complex projects well designed, built to excellent quality, and completed on time.

Case studies

These two case studies were provided by Gerard Daws of NBS Schumann Smith to illustrate how DM works out on a project, and the value it creates. They represent a particular take on life; every project and your situation will be different again.

Hinkley Point C Associated Developments – Ned Barran, Senior Design Manager

Design Management can be an emotive topic at times, owing to the wide-ranging definitions of the role and the fact that it is practised by so many different groups within the design and construction supply chain. Members of the client's team, project management team and the contractor's team all often incorporate the role of 'Design Manager' within their organisational structures.

The client is primarily responsible for the initial 'big idea' that leads into the design brief and signing off the design ideas prepared by the design team. Likewise, the role of Design Management is a scope item that is regularly included as part of the project manager's toolbox – defining deadlines and deliverables, managing the wider project process and making sure that the design team is providing value for money for the client. Down the supply chain, both contractor and subcontractor teams often include the role of Design Manager, which is frequently a more technical role related to design co-ordination.

It is therefore naturally difficult to pin down a correct interpretation of the Design Management function and we believe it would be wrong to impose a hierarchy of importance to any of the definitions. Our team adds another strand to the definition of the Design Management role.

We work within the design team and are appointed solely by architects – managing the design process from within and representing the architects' interests. We have found from experience when working within design teams that some key themes run though each project:

• Managing the client relationship

• Design programme and tracking progress

• Managing the design team

• Keeping on top of planning issues

• Aligning the cost-and-design process.

One project incorporates all of the above themes. Hinkley Point C Associated Developments is a large-scale project providing accommodation and amenity facilities for over 1500 construction workers for the new nuclear reactor at Hinkley Point. The scheme is spread over three campuses and includes park-and-ride sites. The result is one of the largest permanent and semi-permanent modular projects in the UK, providing high-quality accommodation and leisure facilities for a construction workforce and setting the standard for future developments of this kind.

Managing the client relationship

From experience, a poor client relationship can be triggered by something as simple as the lack of a telephone call or of an understanding of what would be delivered.

It is therefore important that expectations are managed and there is clarity on what is delivered and by when. We produced a 'Design Team Procedures' document, which set out the management structure we put in place, including communication routes, reporting structures, etc., for issue to both the client team and the design team. The document encapsulates in writing the Design Management strategy necessary for the success of the design phase. This was significant for obtaining buy-in to our methodologies from the design team, as well as fostering day-to-day interaction throughout the project team.

Design programme and tracking progress

When working within design teams, it is important to share information that is user-friendly and in a format that designers feel comfortable with. We have found from experience that the more traditional Gantt charts sometimes do not reflect the more iterative flow of design. For us, mapping the design process is about understanding the individual tasks, sign-offs and activities that reflect the actual design process itself that suits the designers' needs first rather than the contractor's. Design Programme Workshops were a fitting means to capture this information. Every Monday we relayed to the team the tasks ahead for that week, as well as the unfinished tasks from the past weeks as part of a 'to do' list.

Managing the design team

Our employer, Canaway Fleming Architects, was lead designer rather than lead consultant, with the key design disciplines appointed directly to the client. It is important under such a scenario that some key issues are raised and dealt with. For example, should the design team report progress individually or as a 'single team'? Likewise, have all elements of scope been correctly defined and have the client and project manager been informed well in advance if input from further design disciplines is required? We would always promote a 'single team' ethos – together the team is stronger.

Planning issues

The development of the design was constantly subject to restriction from the planning process, making early design coordination of paramount importance. An added complication was the wide geographical spread of the design team. By establishing a central communication hub we were able to monitor the information flow and alert the various parties when planning conditions were likely to be put in jeopardy. The key factor for success with this issue was our position within the project principals' operations and understanding of the overall project aims and restrictions. From here our ability to advise and manage the design team was greatly enhanced, improving the overall efficiency of the design process and cutting down on abortive work.

10 Stories

Cost issues

Aligning the cost-and-design process is one of the key challenges faced by the project team. The term 'designing to budget' is also one of the most misunderstood and misinterpreted. Designers cannot 'design to budget' in isolation and, contrary to popular belief, do not have the same in-depth understanding of the cost and benchmarking process as a professional cost consultant. Therefore, some key prerequisites need to be in place in order to design to budget. Firstly, the actual budget needs to be understood, as well as an understanding of what is included or excluded, i.e. fit-out, performance-specified works, etc. Likewise, whilst the cost consultant may sometimes align itself more to the client rather than the design team, it is important that they are invited to all key design-team meetings so that they have an understanding of developments to the design. Lastly, there should be an understanding agreed in advance of when costs will be formally updated and the level of detail included in each cost milestone.

Library of Birmingham Design Management – Lucy Freeston, Design Manager

Our team has been working for architects, managing their design teams, for the past decade. During this time it has become apparent that there are many different industry perceptions of what design management is, and that there are many different 'design managers' carrying out many different roles on behalf of many different parties, be they designers, clients or contractors.

We therefore became interested in how our perception of Design Management fits alongside the main industry interpretations; what the key similarities and differences were and what each different party's key objectives and challenges are.

With this in mind, we chose one of our projects to act as a case study to allow reflection upon the above. Owing to the necessary constraints on the length of this study, we have limited the discussion to the pre-construction design management roles.

Library of Birmingham Integrated with the Repertory Theatre

Project Summary

Client Team	
Client	Birmingham City Council
	Birmingham Central Library
	Repertory Theatre
Employer's Agent and Project Manager	Capita Symonds
Cost Manager	Capita Symonds
Design Team	
Architect	Mecanoo International
Structural/ Services and Specialist Engineering	BuroHappold
Theatre Consultant	Theateradvies
Design Management	Davis Langdon
Specification Consultant	Davis Langdon
Contractor	
Main Contractor	Carillion

Form of Contract:

NEC3, Option E

Procurement Route:

2-Stage Design and Build

Programme:

Stage A/B commencement: August 2008

Employers Requirements issued: December 2009

Practical Completion: March 2013

Design Management Roles

In August 2008, following an open design competition, Mecanoo International was appointed to carry out the design of the Library of Birmingham Integrated with the Repertory Theatre project.

Mecanoo, as lead consultant, assembled a multi-disciplinary team including BuroHappold providing Structures, Services and Specialist Engineering design, Theateradvies providing Theatre Consultancy services and NBS Schumann Smith providing Specification Consultancy and Design Management.

It was clear from the outset that in order to make the project a success there would need to be careful management of both the large multi-disciplinary design team (17 disciplines in total) and the relationship between the design team and the client team, which was formed of numerous stakeholders. For this reason Mecanoo appointed NBS Schumann Smith to carry out the role of Design Manager.

Design-team Design Manager

The role carried out by our team was non-technical and involved managing the design process from within the design team, including:

- Understanding what needed to be provided (contractually) and what needed to happen internally for the design team to meet requirements

- Monitoring the design process, looking at programme/change, etc.

- Managing information flows and communication

- Setting up key procedures and documents

- Providing a single point of contact for the client.

The Design Manager made sure that scope and deliverables were understood and clearly relayed to the team and that client expectations were managed by setting out what would be produced at each stage prior to issue.

The Design Manager was also responsible for programme and monitoring progress. This involved not only having an overview of the design process, but also a more in-depth knowledge of the key issues that were affecting design; understanding what the team needed to provide and then drilling down into the tasks that needed to be carried out in order to meet deadlines.

To do this the Design Manager controlled, and regularly circulated, a daily schedule that listed key tasks and meetings. Information was gathered from attendance at design-team and client meetings, interaction with the team and the use of a centralised mailbox to which all project emails were copied. A co-ordinated design programme was also produced with design-team input to identify key interface issues and dependencies. Completion of tasks was then used to track progress.

The Design Manager was also responsible for setting up communication protocols and design-team procedures so that the team were working in a structured manner, lines of responsibility were clear and Mecanoo could be confident that the right people were receiving the right information in a timely manner. In terms of client communication, the Design Manager provided a single point of contact for contractual issues and information distribution. The Design Manager helped to arrange meetings with the client in line with design-team requirements, and also highlighted early in the process when information from the client would be required. A formal Request for Information (RFI) procedure was also set up and managed by our team.

Client Design Manager

It was also important from a client perspective that the design process was carefully managed. Design Management therefore formed part of the role Capita Symonds carried out as Employer's Agent. This involved a different set of tasks from the aforementioned Design Management functions. Key roles included:

- Understanding the client problem and putting together a project scope and programme to solve it

- Choosing the right design team

- Understanding what needed to be provided by the design team throughout the process

- Managing the client decision-making process and change

- Design reviews and sign-offs.

The role began from the inception of the project and the decisions surrounding how to procure a design team. The approach that was adopted was to look for the right 'people' rather than the right 'design', and so great emphasis was placed upon meeting the designers at interview stage, and competition submissions were focused on how the project would be delivered rather than on images.

Capita Symonds also needed to make sure that the client team were set up in a manner that allowed for managed interaction with the design team, so that client requirements could be relayed by the right people at the right time. In order to achieve this, the project was divided into eight sub-projects and a project governance structure was established, recognising distinct building users and numerous stakeholders. In this way it was possible to determine the key people who should attend each meeting and a clear line of responsibility was established, feeding into a set of robust decision-making protocols.

Capita Symonds also provided support for the client in terms of reviewing the design at each stage. It was important that comments were collated in a timely manner and that the implication of comments in terms of change was well understood.

Architectural Design Manager

There was also a requirement for the management of coordination within the design. In contrast to the role of design-team manager, this is a technical position involving internal checks and reviews of drawings and reports.

In this guise, it is the role of the design manager to make sure that the design is co-ordinated both within one's own discipline and also between the other disciplines involved on the project. This co-ordination has to be carried out by all design disciplines and is indeed part of the iterative process of design; however Mecanoo, as lead designer, bore the obligation of ensuring that co-ordination was carried out across the design as a whole.

Conclusions

Although the three roles described above all involved the execution of different tasks, the common thread that runs between them is the goals the managers are driven by:

- Meeting client expectations

- Meeting programme

- Meeting budget

- Producing a quality design.

It is the varying levels of detail in terms of involvement in the design and the point in the process at which the design managers get involved that are the differentiating factors between the roles.

The client Design Manager is primarily responsible for setting the targets and putting together a strategic programme for the project, including design-phase durations and contractual milestones. The client design manager does not need to know what the design team are doing on a day-to-day basis. Rather, they need to know what they will produce, what meetings they require, and then they need regular updates on the progress of design to ensure that it is developing in line with client requirements.

The design-team Design Manager interprets these goals and then breaks down stages and deliverables into tasks that need to be carried out and information that is required. This role requires far greater in-depth knowledge of what the design team do on a day-to-day basis, so that the Design Manager can track where any delays might be occurring and help to mitigate them.

The architectural/engineering Design Manager then makes sure that these tasks are carried out in a co-ordinated manner that meets client expectations and legal requirements. This role involves a detailed understanding of all aspects of the design. It is important that in this role information required is identified early and communicated to the design-team design manager and the client.

11 Future

The CIOB Design Manager's Handbook

Some thoughts on the technology of design, DM, the future of the industry and BIM or – 'Where did the Design Manager go?'

'It's Design Management, Jim, but not as we know it!'

–Mr Spock (with apologies to
Captain James T. Kirk of the Enterprise)

'But what [. . .] is it good for?'

–Engineer at the Advanced Computing Systems Division
of IBM, 1968, commenting on the microchip

Introduction

The baby-boomer generation of which I am one, born post-World War 2 and before 1964, has probably experienced the most change within their lifetime for several generations.

Much of it is technology-driven, but who could have predicted the impact of the microchip in 1968, leading to the internet generation, global connectivity and communication on an unprecedented scale?

In my lifetime mobile phones have become mini-computers in the palm of your hand. If you want to know anything, find anyone, you just look it up on the web. We are being swept along on a tide of technology, impacting our daily lives. Now we are continually looking for the next new thing.

You might wonder why I consider this context important for the Design Manager. The projects we create are very much products of their time – 'the Spirit of the Age or Zeitgeist', as it has been called (Nikolas Pevsner, *Pioneers of Modern Design*). In the design process all these factors, ranging from the global to the intensely local, are synthesised to a lesser or greater extent. DM operates in a connected world, and flourishes on that connectivity and influence.

It is in this context that we produce projects, buildings and physical assets and so the background is key to the DMer, as some of these drivers will be very visible on projects at different times. Examples are: designing for a world where it could be warmer, with more extreme weather conditions, dwindling natural resources, rising energy costs, increasing population pressures and economic uncertainties – and not forgetting drivers that affect the design–construct process, making it more efficient, faster, leaner, etc.

The Design Manager's Handbook, First Edition. John Eynon.
© 2013 The Chartered Institute of Building. Published 2013 by Blackwell Publishing Ltd.

Shift Happens

On the web there is a presentation called 'Did you know? (Shift Happens)', which in 70 or so slides catalogues and presents a series of facts about global change. It is a challenging and thought-provoking presentation and, whether you agree with it or not, it bombards you with the realisation of the pace, impact and immensity of change that is coming at us from all directions. And the pace of change is accelerating!

(Source: www.slideshare.net/lynnmjoy/shift-happens-uk-version, accessed 12/06/2012). See also: http://shifthappens.wikispaces.com/versions

Today, technology is advancing at a rapid rate of knots. This is underlined by the Australian business futurist, Morris Miselowski's quote: 'We've experienced more change in the past 2 years than in the previous 20 years and in the next 10 years we'll experience the equivalent of 100 years of change! Belt up, because we're in for quite a ride!' – www.deliberatepractice.com.au

A while ago there was a list published of 20 jobs that had not been invented yet:

- **Body part maker**: Creates living body parts for athletes and soldiers.

- **Nano-medic**: Nanotechnology advances mean subatomic treatments could transform healthcare.

- **GM or recombinant farmer**: That's 'GM' as in 'genetically modified' or engineered crops and livestock.

- **Elderly wellness consultant**: As an ageing population increases in size, we'll need folks to tend to their physical and mental needs.

- **Memory augmentation surgeon**: Like *Eternal Sunshine of the Spotless Mind*, surgeons could boost patients' memory when it hits capacity.

- **'New science' ethicist**: With the rise of cloning and other ethically dubious practices, ethicists will be needed to ford the river of progress.

- **Space pilots, tour guides and architects**: Space tourism will allow for space pilots, tour guides and the architects that will allow them to live in lunar outposts.

- **Vertical farmers**: The future of farming is straight up. Vertical farms in urban areas could significantly increase food supply.

- **Climate-change reversal specialist**: Regardless of what you think about human-induced climate change, it is clear we'll need scientists who specialise in altering it.

- **Quarantine enforcer**: When a deadly virus spreads rapidly, quarantine enforcers will 'guard the gates'.

- **Weather-modification police**: If weather patterns can be altered and adversely affect other parts of the world, law enforcement will be needed to keep things legal.

- **Virtual lawyer:** As international law grows to supersede national law, lawyers will be needed to handle cases that involve people living in several nations with different laws.

- **Classroom avatar manager**: Intelligent avatars will replace classroom teachers, but the human touch will be needed to properly match teacher to student.

- **Alternative vehicle developers**: Goodbye, internal combustion engine. Zero-emission cars will need smart people to design and manufacture them.

- **Narrowcasters**: As in, the opposite of 'broadcaster.' Media will grow increasingly personalised, and we'll need people to handle all those streams.

- **Waste-data handler**: Think of it as an 'IT axe man' . . . for information. Waste-data handlers will destroy data for security purposes.

- **Virtual clutter organiser**: Now that your electronic life is more cluttered than your physical one, you'll need someone to clean things up – including your e-mail, desktop and user accounts.

- **Time broker/Time bank trader**: What's more valuable than precious metals, stones or cold, hard cash? Your time.

- **Social 'networking' worker**: A social worker for the Web generation.

- **Branding managers**: These already exist for celebrities, but now everyone needs a 'personal brand' so others can easily digest who you are and what you stand for.

> *(Source* – http://www.smartplanet.com/blog/smart-takes/top-20-most-popular-future-jobs-of-2030-vertical-farmer-limb-maker-waste-data-handler-narrowcaster/3384?tag=search-river

> It is a sobering thought to realise that today's younger generation are being educated and trained for some jobs that don't even exist yet, and also that by the time they graduate a significant proportion of everything they have learned will already be obsolete.

Knowledge and skills can always be acquired and learned. We are entering a time when the real required skills are to be able to adapt and roll with change. We need to have the skills and culture of knowledge acquisition and data management, combined with the mental flexibility and agility to see possibilities and potential in new situations.

It is this ability to ride or surf the wave of change that will be critical. DM in a pre-BIM world is backward-looking and output-driven. DM in a post-BIM world will be connected on an unprecedented scale, and knowledge- and information-driven. We are now living through an evolutionary step change in the built-environment industry.

We will look at BIM later in this section, but first we take a look at a few of the waves or tsunamis that will be on all our agendas for the next decade.

Tsunamis

Drivers of change

There is a project by Arup and their Foresight Team called 'Drivers of Change': (www.driversofchange.com).

They have produced a toolkit which looks at these factors under the following headings:

- Energy

- Waste

- Climate change

- Water

- Demographics

- Urbanisation

- Poverty.

In addition these are cross-related to 'STEEP' categories (Social, Technological, Economic, Environmental and Political). The website and toolkit provide resources to make us think about these factors and can be used in a variety of settings, to provoke discussion and action.

The challenge is that so much of this is interconnected and crosses boundaries between subjects and sectors.

Where are the impacts on the construction sector? How will the DMer be affected? Well, I think there are a few headlines.

Climate change and carbon emissions

The pressure to reduce our carbon emissions to mitigate the effects of climate change is increasing. In the UK we already have the statutory Carbon Reduction Commitments that the UK Government is committed to.

The buildings we design now need to be long-life and loose-fit to accept new technologies in their later life, as new renewable technologies are developed. Temperatures in the UK are likely to be on average 2 or 3 degrees warmer!

Perhaps 60 per cent of the buildings existing now will still be operational in 2050 and beyond, so while a lot of our attention is focused on new build, in reality we will need to become much more adept at retro-fit techniques, whether that is simply installing new technologies or a more comprehensive refurbishment and upgrading to new standards.

Energy and resources, waste

Along with the pressures to reduce carbon emissions, we are already seeing moves to reduce energy and resource consumption. Rising prices in the energy market, combined with rising oil costs and reducing supply, will force building owners and operators to look for more efficient built solutions. Reduction in waste in all parts of the process will become more important also.

This will lead to a shift in focus to be as much on operational costs, the whole life cycle, as well as the initial capital expenditure on projects and assets. *(Refer to the Carbon Action 2050 website,* www.carbonaction2050.com *)*

Demographics

- There are now more people in the UK aged 60 and above than there are under 18.

- There are 10.3 million people aged 65 and over.

- Almost 1 in 5 of the population is of state pension age..

- Each year about 650,000 people retire.

- The number of people aged 60 or over in the UK is predicted to rise by more than 50 per cent in the next 25 years.

- By 2083 about one in three people will be over 60.

- The number of people over 85 is expected to double in just two decades.

–UK figures, Sunday Times, 28th January 2012

We're living longer and wanting (or is it needing?!) to work longer. This is putting pressure on our existing models of pension provision and employment. Couple this with the impact of Generation Y – the digital natives, those born post-1982; as they take over leading our businesses they will change the way we work. According to *The Guardian*: 'They care less about salaries, and more about flexible working, time to travel and a better work–life balance.'

They are just more connected, and at home with the technology, as this is now hard-wired in their DNA – for some of us in our later years it has had to be transplanted!

Ironically there is a shortage of younger people entering the construction industry, with many predicting a skills gap (CIOB Report – *Skills in the construction industry*, 2009). At the moment also there is talk of the 'lost generation' not getting employment because of our current economic woes. There may not be enough skilled people available to fill roles in the industry, as an ageing work pool begins to finally retire.

Also there is a crisis in confidence in education and training developing, as graduates can take several years to become proficient and competent in their role.

Integration

Over the last decade or so in the UK, there has also been a trend towards integration of procurement. This has resulted in the complete project team being brought together at the start of projects by customers; in many cases it has been contractor-led, but also been led by project managers and other members of the team. While there are discussions around the quality of solutions that have been delivered, the trend will probably continue, particularly as customers see this route as delivering better value, and certainty earlier in the procurement process. Collaboration and earlier integration will be on the Design Manager's agenda, as they will have a key role in the process. The growing adoption of BIM and IPD models will reinforce this trend.

So the DMer of, say, 2050 will be working in a more energy-conscious, carbon-conscious, connected, integrated and information-reliant and -intensive world. Lifestyle choices, education and training will be very different, and the make-up of our current professional disciplines and roles could have changed beyond recognition.

And then along came BIM . . .

'We're all architects now . . . '

–Rob Charlton, BIM Live, 2010

Introduction

During the course of 2011, Building Information Modelling has become a hot topic and, principally through the initiatives driven by Paul Morrell, the Construction Advisor to the UK Government, and his team including Mark Bew and David Philp, it will continue to be so.

The setting of public-sector BIM targets for 2016, numerous conferences and increasing use of BIM environments on projects are forcing it into the industry's consciousness. Key targets are to reduce capital expenditure by 20 per cent and to meet our carbon reduction targets.

The government-led strategy in the UK is gaining momentum and progress has been made this year in the setting-up of various industry working parties. Between

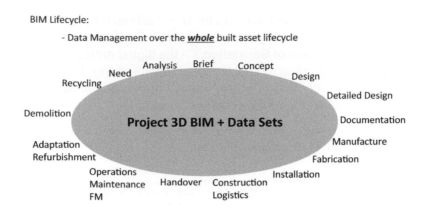

BIM Lifecycle:

- Data Management over the ***whole*** built asset lifecycle

Figure 11.1 The BIM life cycle.

the completion of drafting and the publishing of this Handbook, I expect we will see significant progress on the UK Government agenda, particularly as the results of the working groups begin to be published and take effect, which should happen throughout 2012. This will provide standards, process, protocols and frameworks for implementation and guidance. The USA is a few years ahead of the UK; a significant proportion of projects there are delivered in BIM environments, and the use of Integrated Project Delivery is rising. The benefits of this approach have been well documented. I have highlighted some BIM publications, including the UK strategy, in the bibliography. There is a contrast between the approach to BIM in the UK and that in the USA. In the USA it has been very much bottom-up, and in UK top-down.

> Phil Bernstein, VP for Industry Strategy at Autodesk – '[Distinctive] in the UK . . . is the clarity of objectives and the elegantly designed structure that Paul Morrell has put in place to isolate the key standards problems and get the answers quickly . . . What we have in the States is a lot of messy experimentation under no standards whatsoever.' *(Building Design, 6th January 2012)*

However, industry-wide adoption in the UK will take some time. In the interim the *Innovators** and *Early Adopters** will continue to forge ahead, and within a relatively short period will begin to turn the benefits of a BIM-based approach into significant competitive advantage. Meanwhile, the *Early* and *Late Majority** will continue to debate over how much to adopt and invest in order to catch up. Some will still question the need. There is a danger that a two-speed industry could develop, and for a while this may well happen. There will be those who can see the potential benefits and competitive edge, while others will be more concerned by having to invest and innovate at what is a difficult time for the industry. In addition, while major designers and contractors can afford the time and investment to make the transition, SMEs in the design, construction and supply-chain sectors will find it harder to make the changes required.

The degree to which BIM can be implemented on a project will be affected as much by those partners with the least BIM capability as well as the most! It will be all very well to have a wonderfully data-rich BIM model produced by the design

11 Future

team, only to find it falls off the cliff edge when key subcontractors are incapable of using the model information to manufacture construction elements. And trailing way behind will be the *Laggards** who, having left it so late to come to the party, will probably by this time be in serious danger of going out of business, having been long left behind by their faster, more competitive and more agile competitors. (**Crossing the Chasm* – Geoffrey A Moore; see bibliography).

Implementing BIM is much more than learning a new software suite – it is a different way of working, affecting technology, project process and the culture and roles of the people involved. Most of us will be working for a while, perhaps for several years, in a mixed environment. Some projects will be fully BIM-ed up, others partially so or not at all. This will require considerable flexibility from us all, at all levels and across the industry, in adapting to changing circumstances and developments. However, we mustn't lose sight of the objectives here, particularly that of moving towards a more integrated and intelligent connected industry. As someone remarked to me recently, what is happening is only the industrialisation of the building industry, something that happened in petrochemicals, automotive and aerospace years ago!

For BIM to be effective, it requires more integrated team working and collaboration. This is quite a challenge for some aspects of our industry as it requires a more open, sharing way of working. The insular tribal habits are sometimes hard to resist! – but we need to overcome them in order for BIM to be successful. In addition, the legal and contractual frameworks will also need to flex and catch up, but the Early Adopters are showing this can be achieved, and government is providing the necessary impetus to ensure we all move in the right direction.

Interoperability between platforms is also a key issue. While this is not perfect at the moment with the use of IFCs, it will improve and it does work at present. The UK strategy is insisting that the software developers continue to work together on solutions and hopefully that will provide the resources required, but will also maintain the industry competition necessary to ensure we continue to innovate.

In 2011, with the intervention of Paul Morrell and his team setting the public-sector agenda and targets for 2016, we have reached a turning point – critical mass, even. The tide has turned and it is now only a matter of time in terms of industry adoption of BIM. The question is no longer 'Why?' and 'If', but more 'When?' and 'How much to invest?' The only real question now is, where do you want to be on the Chasm Theory adoption curve? With the Early Adopters or with the Laggards? Your call!

So let's explore BIM in a little more detail. Just remember that this section is intended only as a brief introduction – there are plenty of good handbooks and guides already published. Read the UK Government strategy, get the articles and a few publications, go to some conferences and talk with practitioners to get a feel for what BIM is about and the implications for the design and construction process.

We will look at:

- **Definition and discussion**
- **Aspects of BIM**
- **Implementation**
- **Benefits**
- **Developments.**

BIM Definition

Proposed definition of Building Information Modelling (BIM) as a starting point for discussion and refinement:

> "Building Information Modelling is the digital representation of physical and functional characteristics of a facility creating a shared knowledge resource for information about it and forming a reliable basis for decisions during its life cycle, from earliest conception to demolition."
>
> This definition by CPIc is based closely on the US National BIM Standards Committee (NBIMS).

Discussion

Notice there is no mention of 3D modelling! The 'Building' in BIM might be a built environment asset of some kind, e.g. a piece of infrastructure, a nuclear power station, a rail station, an airport, a suburb of a city even, or simply just a building. The asset has a life cycle, which will follow the timeline of definition, procurement, delivery and operation.

The BIM is the database of information about that asset that potentially can live with it over its entire life cycle – from briefing information, through design, procurement, costing, contract, site delivery and construction, through to completion, handover, and operations and maintenance information to form the basis of FM services. And then it subsequently provides the basis for reuse, adaption and extension, to eventual demolition, removal and recycling.

The information in that database could be in many formats. Yes, there will be 3D design files, say in Revit, Bentley or Tekla, but there will also be spreadsheets, word documents, pdfs and all kinds of files containing technical, programme, cost information and any other information that can be attached to the database.

Current thinking seems to be moving to the concept of having a relatively small model file, linked to a database in SQL or something similar, which holds the majority of the information. Naturally this information can be catalogued, searched, queried and output however you need or want it to be.

So while many people equate BIM merely with 3D modelling, it is in actual fact much more to do with *data management over the asset life cycle*. Picture all the information that is collected and circulated over the lifetime of a project – BIM opens up a world where that information and knowledge can be truly integrated and searchable. Yes, of course much of this information will be in 3D format, but there will be so much more as well.

At this point it is worth mentioning COBie – the Construction Operations Building information exchange. This is a specification or format for the capture and delivery of information needed by facility managers over the asset life cycle. COBie information is produced at all stages of the project and currently is usually presented as an extensive spreadsheet. Part of the UK strategy is for public-sector departments to determine their required COBie drops or data exchanges at each of the project stages. These will be published as part of the BIM process work currently under way, and due for publication in mid-2012.

Aspects of BIM

Let's consider how BIM can be used on a project or asset. I'm trying to make this reasonably neutral in terms of mentioning particular software or packages or companies facilitating BIM process or activities.

3D Design co-ordination and clash detection

This is probably one of the best-known aspects. Taking each of the discipline design models, it is possible to aggregate the models into one environment and identify clashes and conflicts between the discipline designs. Whether this is one totally integrated model or a 'federated' model composed of several models is a discussion. Certainly the federated approach reduces file sizes and lessens the impact on ICT infrastructure, which is a big consideration in implementation. It is possible to federate models produced by different software, through use of a package such as Navisworks, which will enable clash detection.

It is also possible to use IFCs to import a model into another package. I've heard varying reports on how successful this can be, but certainly the future will be about more interoperability between platforms.

4D Timeline/programming

This is the generally accepted '4D'. It is possible to assign time attributes to parts of the model so that a view can be generated which shows the construction sequence. Into this can be added elements such as cranes, hoardings, scaffolding, hoists, etc. Most of us have probably seen these animations. Generally they are used to explain the construction sequence and site logistics to, say, a client, site team or supply chain. This is particularly useful for understanding the requirements for the construction sequence and the implications in 3D.

5D Quantity take-off and cost planning

Once the model is designed, even in outline, it is possible to take off quantities for elements or components. The output can be in a form which cost-planning or estimating software can accept, although how this actually works needs to be defined up front to make the technical aspects feasible. Over time, as libraries of project-based data are established, this will enable information to be accessed to compile cost plans quickly and efficiently. It is worth noting that the use of libraries of information to populate the model and database is a great factor in improving efficiency, speed and reliability of the project process.

Simulations – lighting, fire, people movement, thermal, carbon, energy

In addition to the basic 3D design packages, there are additional plug-ins and various packages that can use the model to simulate various conditions such as lighting levels, fire performance, the movement of people, including lift simulations and queuing patterns, thermal losses and transmittance, carbon emissions and energy consumption. In compiling the model, the relevant data will need to be input as attributes to elements or components to enable the functionality required. But again, as libraries of data are developed, then the input activity can be accelerated.

Operations + Maintenance manuals and information

The project BIM can be used as a means of storing and accessing O+M and CDM Health and Safety File information. This can be made available to a hand-held tablet, so that by going into a room you could call up the room data sheets, relevant manufacturer's information and so on. If, say, an item of plant or FFE has a barcode

11 Future

or an RFID tag, then it can be scanned and used to access the information on that item, such as dates of manufacture, installation, defects period, maintenance and any other related information you care to choose to have stored.

Visualisations

Another aspect that is probably familiar to most people is visualisations of the design model, utilising stills from selected 'camera' views or animated fly-throughs around the site and building. These are particularly useful in explaining the scheme to customers or stakeholders, and also for explaining the scheme to the planning department or other interested parties. It is possible to have a fully virtual immersive experience, with headsets, goggles and gloves, so that you experience the model from inside. Most of the BIM-enabled universities, such as Salford for example, have this kind of facility. These kinds of animation are extended into the construction sequence animation discussed above.

Site safety planning

Some contractors use the phrase 'Build it twice' – i.e. build the project virtually, and then go and physically build it. As the BIM model becomes more detailed and contains the construction status information, then it is possible to review the construction process in detail from a safety perspective, considering sequences, access requirements, and particular areas that may require special temporary works or other factors. This takes risk assessment to a new level, enabling teams to view how to construct virtually, before going out into the field to actually do it.

Fittings, fixtures and equipment

Having designed the 'base-build', it is equally straightforward to design in FFE requirements, link these to a room data-sheet package such as Codebook, and link to procurement, O+M information, etc. Manufacturers are beginning to produce libraries of their product ranges, which will make it easier for designers to incorporate them in their models. Also there are library sites being developed, such as NBS BIM library and the BIM store.

Progress management

Using the timeline and logistics elements, it is possible to translate the proposed programme of construction events into an animated view. Site managers using tablets can record progress and problems and these can be uploaded to the model to give a model view of progress.

There is an interesting study called KanBIM, which looks at BIM-based progress combined with lean thinking: see www.tekla.com/international/about-us/news/pages/bims-return-on-investment-wheres-the-beef-now.aspx.

Off-site manufacture

Having developed a detailed, clash-free data model, subcontractors can retrieve the information they need to manufacture their elements and components. In some cases the data can be exported straight to CNC-enabled plants, which enable computer-generated manufacturing processes. Double or triple handling of information is avoided, and information is taken directly from the model to facilitate the process. As supply chains develop object libraries of their systems and components, these can be made available to designers to incorporate within the design-stage BIM. Of course, with the physical data other attributes can be added, such as procurement data, costings, lead-in times, technical performance data to enable simulations, and O+M information.

Life-cycle costing and management

Using plug-ins or additional software packages it is possible to run simulations of the life cycle, looking at carbon emissions, energy consumption and so on. Therefore from an early point in the design process, decisions can be made from an operational perspective to develop the optimum solutions.

It is a short jump from there to:

Facilities management/Building operations

The project BIM is updated to as-built at construction completion. Loaded with the O+M information, this can then be used by the customer's FM team to manage the asset. This aspect is part of the UK Government's drive on BIM adoption for the industry. If consistency of data is applied across the public-sector estate, then this can be harnessed to result in increased efficiency and management.

As discussed earlier, the BIM can be accessed using hand-held tablets, so that managers walking round the asset can retrieve and also update information. The use of barcode readers and scanners can enable quick access to relevant information on particular items of plant or equipment. The use of tablets has been extended into snagging and defect logging, following similar principles.

Recycling

At eventual demolition and/or recycling, the BIM can contain the information to enable safe demolition and also the material specifications for recycling if required.

ArtrA Pilot Project – Basic Principles

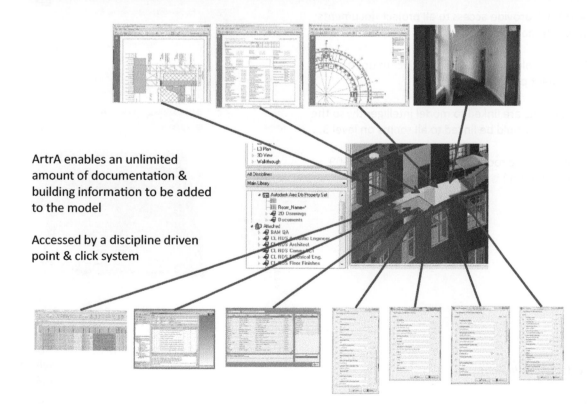

ArtrA enables an unlimited amount of documentation & building information to be added to the model

Accessed by a discipline driven point & click system

Figure 11.2 ArtrA. Reproduced by permission of ArtrA Ltd. (www.artra.co.uk).

11 Future

ArtrA Pilot Project - Linking Databases & Models

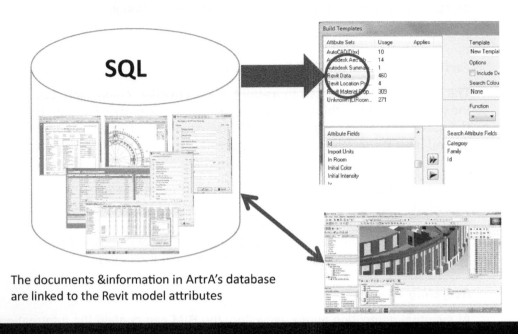

The documents &information in ArtrA's database are linked to the Revit model attributes

ArtrA Pilot Project - Documentation

ArtrA will provide access to all project documents through the model

The principle is to click on an object to open any document pertaining to it

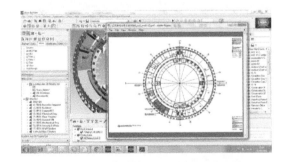

Documents are linked to model intelligently, so the floor plan would be linked to all rooms on level 3...

...whereas the room data sheet for Room 03-09 (Conservation) is linked to that room only

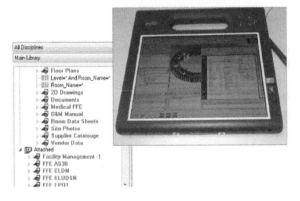

Figure 11.2 (*Continued*)

Batches of documents can be linked to any object(s) in the model for near instant review in the office and on site...

In this example a PDF of the Third Floor Proposed GA Plan will be linked to all rooms/spaces on the 3rd floor

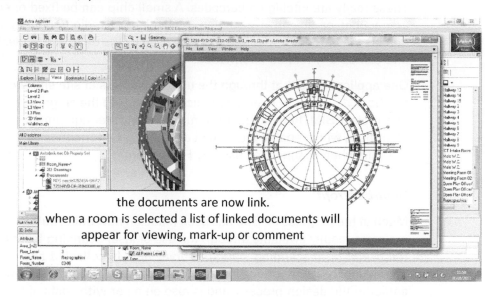

the documents are now link.
when a room is selected a list of linked documents will appear for viewing, mark-up or comment

26

ArtrA Pilot Project – Using Barcodes for Facilities Management

Modern tablet PC's are equipped with barcode & RFID readers

ArtrA uses this technology to help the construction team and facilities managers navigate the model & find information

E.g. When a barcode on a door jamb is scanned, ArtrA zooms the model to the correct place and lists all available documentation & information Tags

11 Future

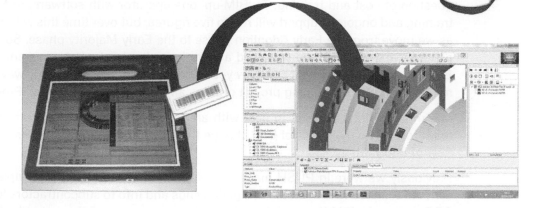

2

Figure 11.2 (*Continued*)

In addition, assuming the BIM is kept up to date, then should alterations or extensions to the asset be proposed, the BIM provides a good starting point for the new design team.

RFID (Radio-frequency Identity tag)

These really are intelligent barcodes. A small chip can be fixed or embedded in a component or object; it can then be read by a radio-frequency reader, which detects the barcode number, and can be used by the database to identify the object and call up any related information.

One application is that through the use of scanners at site entrances, contractors can track deliveries of materials, which could link to the progress-tracking element of a project BIM. Items of plant or furniture or other equipment can have unique identifiers, which enable the correct and relevant information to be called up from the BIM when scanned, as used by FM managers.

Refurb/retrofit

Much of the progress on BIM has been made on new build, but now survey models are more readily available through the use of laser-scanning technology. Through the use of laser-scanning stations, survey data is collected and converted into a 3D model to be imported into the BIM model. This provides greater accuracy earlier in the design process, and is also on a par with costs of traditional survey methods.

Incidentally, the same technology is being used to scan road-accident scenes, speeding up the process of reopening the roads to traffic.

Implementation

As I have highlighted, moving into a BIM environment is not an overnight step but more of a journey. The UK strategy is very realistic, both in the timescales and the targets set, as not being overly ambitious by 2016, giving time for industry to respond in a more coherent and co-ordinated way. However, for the Innovators and Early Adopters, and some UK pilot projects, 2014 is a more realistic deadline.

The technology aspect is perhaps the easiest part in that the software is developing all the time, and capabilities are increasing almost on a daily basis. It is a question of cost and training. To 'BIM-up' one operator with software, hardware, training and ongoing support will run to five figures, but over time this will reduce, as we move from the Early Adoption phase to the Early Majority phase. So this is a significant exercise and you need a plan, to understand what you want out of BIM, why you want to do it and how you're going to get there. In many ways, it is not dissimilar to developing project process as discussed in Section 2: Process.

The more difficult aspects to deal with are intertwined in process, culture, and team roles. The dynamics of BIM change the way project process works and the flow of information.

Previously we were in PUSH mode: as information is produced, the outputs are sent out to those who need them, e.g. drawings and info to subcontractors to start CDP design. In BIM the model is effectively the output, and if you need information you go and get it, when you need it, and define how you want it, format presentation, etc. The responsibility shifts to PULL – those that need information for their activities get it from the source: the project BIM. This changes the dynamics, and the DMer/BIM Manager/Data Manager changes also.

Benefits

The benefits of working in BIM are well catalogued. The BSI Investors Report provides a good snapshot of key issues. The BIM Handbook catalogues a number of detailed case studies. (Both are referenced in the Bibliography.)

BIM offers a number of benefits, including reducing the rework of design and site works, resulting in more certainty in the design and procurement process and at an earlier stage. This enables more efficient ways of working, as well as producing more viable design solutions that are value-managed closer to a customer's requirements from the start. Significant time and cost savings can result over the asset life cycle, as well as in just the design/procure stages.

In discussions with designers, they're open about producing better work, more quickly and using fewer resources. A contractor publicised recently that for a modest investment on BIM on a new project, they had more than recouped their investment in a few months, through clash detection and avoidance, and risk elimination.

Developments

So between writing this in February 2012 and publishing towards the end of the year, I am expecting a lot to happen in this arena. Certainly the UK Government agenda will kick in, and we will see outputs from the various working parties.

These will include further parts of BS 1192 (Parts 2 and 3) relating to BIM for Capex and Opex, process mapping, looking at the various Plans of Work, and harmonising these with the COBie outputs required by the public-sector groups. The outcome will be industry-agreed standards, protocols and classifications which then provide a firm basis for moving forward.

Further industry groups will develop in response to various parts of the agenda. I also expect more focus on SMEs and the trailing edge of the industry, to get them to be BIM-intelligent.

We will continue to see case studies and pilot projects publicised by the leading BIM players to help educate about what BIM is and the benefits it can bring.

Technologically, in society at large we are seeing a move towards more mobile, portable and connected solutions – so the use of tablets like the iPad, mobile computing using the Cloud for data and software is going to rise. In BIM-world things will be no different, and the prospect of carrying the BIM round with you on site, using a local wireless network on your iPad, is in fact already here. It is just going to get better and more common in usage.

in the longer term, who knows where this can go? We have discussed the idea of the 'industrialisation of the construction industry'.

Imagine you're a client who needs a new office. You go to a showroom and sit with a 'visualiser', who shows you various designs and ideas. She takes a brief from you, works up a solution as a BIM from libraries of information, and identifies costs, programme, supply chain, etc.

You even pick the furniture blinds and colours. As a final sign-off you take a tour of the model in a headset, accessing the model as an immersive environment. You can see people at work, hear the sounds, smell the coffee. Planning permission and building-code compliance is completed online through tapping into the local authority urban BIM. You sign the contract, production slots are booked online and you book in the technical team to construct the office when it arrives in containers on your site.

Alternatively you want a new house. You go to amazon.co.uk . . . you can imagine the rest . . . it arrives on a lorry a few weeks later!

Why not? We already have the technology!

The beauty of the BIM approach is that it answers many questions from over the years, such as integrating the industry, more collaboration, a less adversarial culture, lean design and construction and so on. This follows on naturally from the legacy of Latham and Egan. Also it opens up a whole new world in terms of the knowledge and power that is potentially at your fingertips. Design, cost, logistics, simulations, FM – it can all be right there. Knowledge is power, and BIM opens up the connectivity of asset information across the life cycle.

As Rob Charlton said at BIMLive 2011, 'We're all architects now.' But equally, we could all be contractors, QSs, project managers and engineers to some extent. With the right training and expertise, combined with the right BIM models, libraries and background data, the possibilities are opening up and we all have those opportunities, whatever our current role or discipline. Admittedly it is going to take several years to reach this position, but the scene is set for new players to enter our industry, in some cases from unexpected directions.

Expect a blurring of roles and perhaps new roles appearing. This will ripple down into the way industry professionals are educated and trained – eventually perhaps affecting the composition of our institutions. Those that embrace collaboration and connectivity, rather than insular pragmatism, will flourish. This is a time when our fragmented industry can truly come together.

This brings us back to the DMer. The future of the Design Manager is entwined with the future of BIM. Management of the model, management of the *data*, will be the keys to project success.

At a really high level, perhaps DM is really fundamentally about information management – managing the activities and outputs so that project stakeholders can get the right information, at the right time, at the right cost and quality. This leaves DMers ideally placed to move into the role that ensures the management of data over the complete asset life cycle – *'Kings of BIM'* – Discuss!

Afterword

So that is it, for now anyway. I hope that you will find this edition of The CIOB Design Manager's Handbook useful.

There can always be a discussion around the level of detail. I see little point in regurgitating stuff that others have produced more intelligently and eruditely; I would rather point you in their direction. The bibliography and appendices will give you further resources and ideas for reflection.

All that remains is to say:

- Join us on LinkedIn at the Design Managers' Forum, and on the CIOB website to learn about the current developments in DM.

- This is a really exciting, challenging and developing time for everyone involved in DM.

- Take the opportunity to post on the Forum, or contact us with any feedback, ideas or comments.

- I wish you well in your journey in DM!

And finally for this Handbook (!)

It is time our industry grew up a little and got beyond the juvenile, tribal picture of life that we have and that is hard-wired into our professional perceptions.

Whether designer, supplier or contractor, architect, technologist, engineer or surveyor, we need to lose this 'us and them' mentality – no one discipline holds the answer to the issues that confront us all.

It is together that we are the solution, collaborating and integrating, moving forward in innovation, in a culture that joins the dots – that *connects*. What is needed is an integrator, a driver for connectivity.

Strangely, whatever your persuasion or background, this is exactly where the Design Manager fits in!

Addendum – October 2012

Whilst completing the main text of the Handbook in January, I was aware that particularly the BIM agenda in the UK would continue to develop throughout the year – and so it has! In order to try and give an overview of progress, I've added this Addendum to give currency to the handbook and some headlines on current developments.

The headline here is that our industry is and will be changing in an unprecedented way, and these changes will affect everyone, and also DMers and the way in which they work.

BIS BIM strategy

The working groups have continued throughout the year, and there have been some notable developments.

Plans of Work – a unified plan of work for the industry is becoming a reality, developed from the CIC Scope of Services and the APM plans. The RIBA Plan of Work will undergo a major rewrite next year, the RIBA having already produced the BIM Overlay as an interim solution, this year.

This is potentially a game changer. If we can have one plan of work that everyone understands and accepts this has to help in bringing the industry together, aiding collaboration, and improving communication.

BS1192 – is being developed and two further documents PAS 1192 Parts 2 and 3, Building Information Management – Information requirements for the capital delivery phase of construction projects, and also for the operational phase, will be published. (Capex and Opex respectively).These documents develop the thinking of BS1192 2007 in data management in a BIM environment. PAS 1192 Part 2 is due to be published in December 2012, the draft having been out for consultation over the last few months. Part 3 will follow on later in 2013.

COBie UK 2012 – is now available on the BIS BIM website, setting out guidance and templates for COBie in the UK, aligned to the data drops that have been defined at the different stages of model development.

Work with various government departments has also continued. The Ministry of Justice has been at the forefront of the process and several pilot projects are now underway, with tender information and requirements being issued in BIM formats, and also the returns for evaluation. BIM implementation is now spreading across other public sector departments.

Initial learnings from the process have already identified substantial savings by involving the FM operator in the design process, leading to efficiencies both in the design, construction and potentially the operational phases. Case studies will be published that capture and communicate these learnings.

It should not be underestimated the impact that the Government thrust will have. Aligning procurement and tendering processes and requirements across public sector departments will in time have a massive effect. Already the top down implementation of BIM in the UK has won international recognition and in time will lead to common standards and processes that will feed into the private sector. In fact, this is happening already with some leading retailers and developers.

Infrastructure – Another aspect has been burgeoning growth of BIM use on infrastructure projects. The London Crossrail project being a leading example but there are now many others. Whilst BIM is perhaps a relatively universally understood acronym, it is about much more than buildings. BIM can be applied to any built asset, enabling data management over its lifecycle.

Growing collaboration – An encouraging sign has been the growth of various groups looking at the adoption of BIM and the desire to work across the tribal silos. We have a long way to go but I think this bodes well as we journey towards a more integrated digitally based industry.

The DMer and BIM – over this year the discussion has continued about whose model is it and who manages it? Client? Designer? Contractor?

I'm not sure a definitive answer has emerged, and may not, but already the concept of BIM Manager, Model Manager or Coordinator is emerging. Through their knowledge of process, design and the delivery phase, DMers are ideally placed to move into this field (If they want to!). It will require gaining a working technical understanding of BIM but the opportunity is there for those that want to develop in this direction. Certainly as use of BIM develops the need for coordination and integration of the process and deliverables is paramount.

It remains to be seen regarding ownership. I suspect it will be difficult to generalise and will vary from one project to another.

Overview – It seems to me from recent meetings and events, that the majority of the industry in the UK remains sceptical about BIM and its implementation. At the moment at a time of economic stringency all investment comes under intense scrutiny. The return on investment discussion in taking this forward is key, particularly for SMEs such as the smaller designers, contractors and subcontractors.

I think in the coming months this will be the battleground for moving through into main stream adoption, in the ability of SMEs across the UK and at all levels and roles across the industry to connect with BIM and use it to move forward.

Whether we've reached critical mass yet it's hard to say but every week we see more reports of major contractors or clients committing to the use of BIM on their projects.

Soon the use of BIM will be the accepted norm, at which point to some extent the ROI argument begins to take second place to simple competitive advantage in the market place. It might become a question of survival for some businesses – innovate or be left behind.

And finally . . .

We held the Design Management conference in October 2012, and we reviewed DM from a number of perspectives including Client, Designer and Contractor. There was considerable enthusiasm for DM and a number of ideas about how we should move forward, particularly looking at education and training, compliance and standards. Again there was no shortage of ideas and plenty of opinions!

As we said at the conference, we have begun the conversation regarding Design Management. So join with us, and let's continue the journey together!

Appendices

APPENDIX A   Lean Project Delivery – innovation in integrated design & delivery

Alan Mossman, Glenn Ballard & Christine Pasquire[1] (Private communication, 2010). Reproduced by permission of Alan Mossman, The Change Business Ltd.

Abstract

Purpose: To review the state-of-the-art of Lean Project Delivery (LPD), to show the relationship between LPD and integrated design and delivery (particularly the IDDS element 'collaborative processes') and to propose further stages of development, research and practice.

Method: Description and analysis of action research & learning in current practice.

Findings: In the context of Lean Project Delivery with Target Value Design (TVD) projects are completed below market cost—so far as much as 19% below and expected cost falls as design and construction progresses.

Limitations: this work is based on a limited number of linked cases in the US over the last 10 years. We suggest the ideas are applicable far more widely.

Implications: there is still more development and research required to develop effective leadership models for integrated design and delivery, create *whole of life* Target Value Design, to create significant bodies of evidence to guide the design of many building types and to further develop the Lean Project Delivery System.

Value for practitioners: initial indications are that the methods described work together to deliver significant benefits for owners/clients/end-users and create a more satisfying experience for most designers and constructors.

Keywords: Target Value Design, lean construction, lean project delivery, allowable cost, target cost, value, evidence-based design, A3, set-based design, collaboration, early constructor involvement, integrated form of agreement, integrated project delivery, value management, IDDS

Introduction

The idea of integrated design and delivery is not new – the UK Emmerson Report of 1962 noted that '*in no other industry was the responsibility for design so far*

[1] Corresponding author: Alan Mossman, Dept of Civil and Building Engineering, Loughborough University, LE11 3TU & The Change Business Ltd 19 Piccadilly Stroud GL5 1HB UK +44 1453 765611 a.mossman@lboro.ac.uk www.thechangebusiness.co.uk

Dr Glenn Ballard, Project Production Systems Laboratory (P2SL), University of California at Berkeley, Berkeley, CA 94720-1712, United States glennballard@mac.com

Dr Christine Pasquire, Dept of Civil and Building Engineering, Loughborough University, LE11 3TU UK +44 1509 222895 C.L.Pasquire@lboro.ac.uk www.construct-lean.org

Portions of this paper will appear in *Lean Construction: People, processes and Production* a forthcoming book by Alan Mossman and Tariq Abdelhamid.

Appendix A

removed from the responsibility for production'. While there has been a shift since then toward more integrated procurement of construction, it has been piecemeal, partial and is still far from the norm, particularly in public sector design and construction.

Many in the public sector believe, usually erroneously, that public procurement rules outlaw integrated design and construction procurement while others actively pursue integrated procurement paths such as PPP & PFI[2]. As the Economist noted recently *'Conventional procurement has too often been a litany of overruns and delays, and it does not create an incentive for contractors to consider maintenance costs. If the PFI brings in construction on budget and time, and upkeep is cheaper, then it is likely to offer value for money'* (Economist 2010)

Others see technology as a barrier (e.g. inter-operability problems) or a solution (e.g. clash detection and virtual construction) to integrated design & delivery (IDD). In a recent issue of *AEC Bytes*, a blog, (7 Apr 2010) Randy Deutsch reminded his readers of GSA's Charles Hardy's statement 'BIM is about 10% technology and 90% sociology'. Deutsch went on to assert *'ninety percent of what has been written, analyzed and studied about BIM so far is the technology. While the 10% technology works itself out,'* he continued, *'we would as an industry do well to turn our attention toward the 90% that we share, the sociology of Integrated Design.'*

The two pillars of the Toyota Production System are illustrated in Figure 1. One is about process and the other about people. Technology is implicit. Of course technology is important to Toyota – what is significant here is that it is always in service to people and to the processes that enable them to deliver value to their customers.

Notwithstanding the efforts of its leadership, discussions within the CIB[3] Priority Theme *'Improving construction and use through Integrated Design and Delivery Solutions'* community have been heavy on technology and light on people and process. In the CIB Whitepaper on *Integrated Design and Delivery Solutions* (IDDS) (Owen ed 2009) they (IDDS) are defined as *'[using] collaborative processes and advanced skills with integrated data, information and knowledge management to minimise structural and process inefficiencies and to enhance the value delivered during design, build and operation – and across projects.'* The integrative nature of IDDS is summarised in Figure 2:

Our purpose in this paper is to describe *action research* on a number of related and integrative **collaborative processes** that we believe enable teams using Building Information Modelling (BIM) and *virtual construction* to integrate design and delivery of projects. The principal processes are:

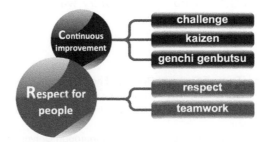

Figure 1 The two pillars of the Toyota Way – *kaizen* = improvement; *genchi genbutsu* = go see for yourself (source: Toyota 2001. Graphic: Alan Mossman. Reproduced by permission of the authors.)

[2] Public Private Partnership & Private Finance Initiative (a subset of PPP)

[3] CIB is the acronym of the abbreviated French (former) name: "Conseil International du Bâtiment" —the abbreviation remains but in 1998 the full name changed to: International Council for Research and Innovation in Building and Construction

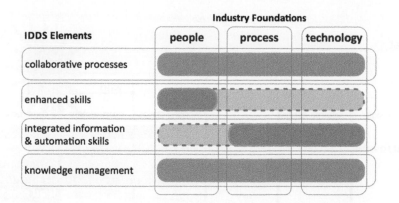

Figure 2 Interaction between IDDS elements and people, process and technology indesign and construction. Darker areas indicate a stronger interaction. After Owen ed 2009. Graphic: Alan Mossman. Reproduced by permission of the authors.

- Lean Project Delivery
- Evidence-based Design
- Set-based Design
- Target Value Design

Lean project delivery emerged in the 1990s and the other three areas are more recent. Target Value Design is a method used within Lean Project Delivery. Evidence-based design also belongs to Lean Project Delivery, and, like Set-based Design, is more a strategy than a method. It is our contention that they all enable integrated design and delivery. Other processes are mentioned in the paper and issues relevant to the skills and knowledge management elements are noted in passing.

Just as invention of the aerofoil enabled the development of aerodynamic theory (Alexander 1974) we believe that the *practice* of lean and integrated project design and delivery is advancing ahead of scholastic research and that *crucial experiments* are to be found in practice. Sociological, ethnographic and psychological studies of these design and delivery processes will help develop our understanding of how people interact with technologies that are integral to those processes and how the industry can make a step- change in productivity.

Why is integrated project design and delivery important?

Figure 3 provides a high level view of the design-bid-build process (top) and an integrated delivery process below. In the top process constructors don't come aboard until the design is substantially complete – then the 'explosion' as constructors struggle to make sense of the design, make it buildable and try to compensate for its limitations. The vertically shaded background represents the extent to which the whole team understands what the client wants and how the project will deliver it.

By contrast, in integrated design & delivery processes constructors join the team at or very soon after the start, they develop their understanding of client need and how it will be satisfied with the designers and are able to develop a cost-effective production process alongside the design. Potential benefits of this approach are summarised in table 1.

We explain how these benefits arise in the rest of the paper.

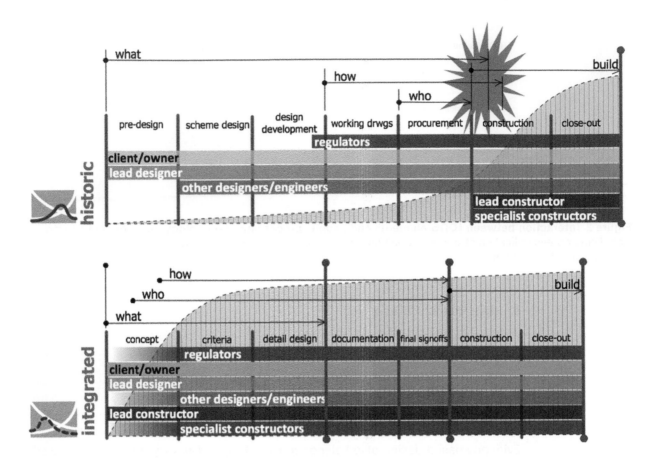

Figure 3 Comparison of historic and integrated project delivery timelines (after Eckblad *et al* 2007) **& their impact on the development of a shared understanding of theproject by the whole team** – vertical hatching (Lichtig 2007). Lichtig suggests that *shared* understanding may never reach 100% in the historic approach as users often find the completed facility different from what they wanted or expected. The integrated model is intentionally shorter than the historic one as that tends to be what happens. The two small graphs to the left of each diagram are *'MacLeamy Curves'* (CURT 2004. Graphic: Alan Mossman. Reproduced by permission of the authors.)

Table 1 Potential benefits of Integrated, Lean Project Design & Delivery.

For clients	• Easier to link design options to business objectives
	• Improved value and a higher quality product
	• Greater potential for lower cost construction & operation
	• Reduced energy cost of use
	• Facility delivered faster with higher quality so able to begin payback sonner
For designers	• Less rework, minimises iteration
	• Relationships, conversations & commitments are managed
	• Decisions at last responsible moment
	• Easier to create excellent *green* buildings
	• Easier to design to target cost
	• reduced design documentation time
For constructors	• Better integrated design → less rework, lower costs, faster completion
	• More buildable, logistics considered from outset
	• Relationships, conversations & commitments systematically managed
	• greater construction process reliability and cost certainty

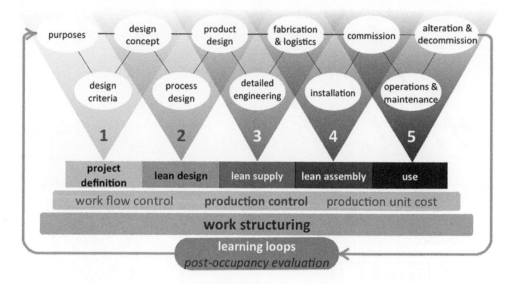

Figure 4 The Lean Project Delivery System™4, **LPDS** (Ballard 2000. Graphic: Alan Mossman. Reproduced by permission of the authors.)

Lean Project Delivery

Lean project delivery builds cooperation in the context of a single integrated team involving the owner, architect, constructor and other critical players as *equals* in the pursuit of a shared goal.

Figure 4 shows the Lean Project Delivery System model. Designed to support a new and better way to design and build capital facilities, it captures both the linear and the iterative nature of the design and construction process and recognises the importance of certain aspects of design and construction happening in parallel rather than sequentially. A post-occupancy evaluation module links the end of one project to the beginning of subsequent ones. (Each element is briefly explained in Ballard 2000; see also Ballard & Howell 2003a).

Value

Value is the *raison d'etre* for the lean project delivery process and it is in large measure what distinguishes lean project delivery from historic delivery processes. It is the client, or more usually the *client system*[5], that defines value. It will be different for each client system and each project. Value can be expressed in a variety of ways – for a university it might include some combination of student and faculty experience, flexibility to allow for changes in research projects and technologies – represented by circle D in Figure 5.

In Figure 5 each circle is in proportion to the relative price, or return, so that if the initial price of design is 1 unit and construction is 10 the price of maintaining the building for 20 years will be between 30 & 50 units, the users operating costs – heating, cooling, salaries, etc – will amount to between 150-300 units. Finally D, the economic and other benefits delivered by those who work in the building will

---

[4] Lean Project Delivery System is a trademark of the Lean Construction Institute www.leanconstruction.org

[5] client 'system' includes the end-users, those who will approve payment, the technical buyers (construction specialists) and their advisors. For many projects, particularly larger ones, government agencies and neighbours often have an input to the definition of value. (Salvatierra-Garrido et al 2009)

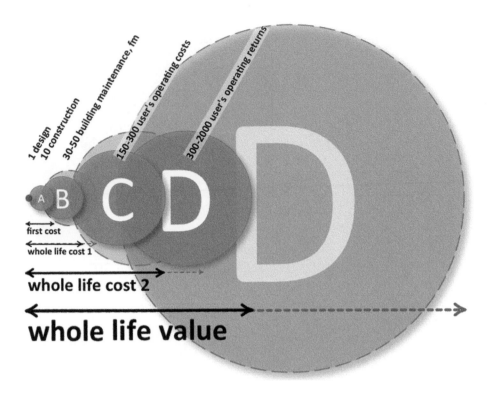

Figure 5 Understanding value in design: Output value (D) in relation to first cost (A+design cost) and whole life cost (A+B or A+B+C) – Diagram based on an idea from Don Ward, Constructing Excellence & Anne King, BSRIA, data from Evans *et al* (1998), Hughes *et al* (2004), Ive (2006) and other. Graphic: Alan Mossman. Reproduced by permission of the authors.

Table 2 Healthcare and educational outcomes.

HEALTHCARE OUTCOM[6]	K-12 SCHOOL OUTCOMES
• Clinical outcomes	• Less truanting
• Hospital-acquired infection rates	• Better test scores & exam results
• Safety outcomes	• Improved behaviour & social skills
• Medication error rates	• Staff retention
• Medication rates	• Parental involvement
• Re-hospitalisation rates	• Employer recognition
• Length of stays	• Community integration
• Patient transfers	• Economic regeneration
• Costs per unit of service	• Community use
• Patient & Visitor satisfaction	
• Staff morale & Staff turnover	

be between 300 & 2000 units. There is a range because the actual numbers depend on the uses to which the building is put – a school or a hospital will have a different return to say a factory or an office – and it is suggested that an office in the City of London or in Manhattan will differ from an office in the suburbs.

The purpose of design is to create a structure or building that enables whatever is in D (in healthcare and in education for example it includes the items listed in Table 2) – yet how often do architects get to know what D is, let alone have the information they need to design to optimize it – the latter is the role of evidence-based design (EBD), the subject of a later section.

---

[6] source: National Health Service Estates (UK)

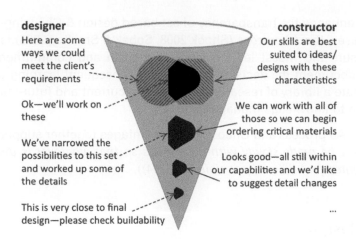

designer
Here are some ways we could meet the client's requirements

Ok—we'll work on these

We've narrowed the possibilities to this set and worked up some of the details

This is very close to final design—please check buildability

constructor
Our skills are best suited to ideas/designs with these characteristics

We can work with all of those so we can begin ordering critical materials

Looks good—all still within our capabilities and we'd like to suggest detail changes

...

Figure 6 Set-based design dialogue (after Sobek et al 1999 Graphic: Alan Mossman. Reproduced by permission of the authors.)

Set-based design

Set-based design (Ward *et al* 1995, Sobek *et al* 1999, Kennedy 2004, Morgan & Liker 2006, Ward 2007) enables a range of discipline specialists, including constructors, to develop a *set*[7] of possible solutions to product design and production design problems and then to decide at the *last responsible moment*[8]. Deciding at the last responsible moment allows the project team time to develop a number of design options in parallel and then choose between them with agreement among stakeholders. All of which reduces the need for later rework.

Figure 6 illustrates the kind of dialogue that might happen between designers and constructors. In practice there are many more than two parties around the funnel – architects, structural engineers, services engineers, façade engineers as well as specialist constructors and the lead constructor. Each has a point of view that can contribute to the optimisation of the project as a whole. These are some of the stakeholders that need to agree.

Faced with complex decisions, often on a daily basis, choosing an effective decision- making method is important as **methods → decisions → actions → results.** Defined by civil engineer Jim Suhr, *Choosing by Advantages* (1999) is a simple, easy to use system that can help individuals and teams make good decisions consistently and show how the decision was arrived at – it includes an audit trail (for an example see Parrish & Tommelein 2009). The system helps avoid common mistakes in decision making – e.g. double counting – by focusing only on the valued advantages of the alternatives[9].

---

[7] what is critical here is the development in parallel of a number of design options – it is sometimes called *concurrent engineering* – in contrast to point based design where a number of options may be considered but only one is developed at a time.

[8] The *last responsible moment* is an important if subjective concept best identified through a collaborative planning process and then regularly reconfirmed.

[9] unlike the *Analytic Hierarchy Process*, a currently supported method in academia, *Choosing by Advantages* does not weight factors or values. Refusing to weight values goes hand-in-glove with engaging stakeholders in the decision-making process. Their values all have equal relevance. The focus on the advantages of alternatives provides an environment in which qualitative valuation can be included, unlike return-on- investment calculations, which require that all valuations be monetizable. *Choosing by Advantages* is used by the Society of American Value Engineers (SAVE)

To enhance the transparency of set-based design & Choosing-by-Advantages processes, the *A3 process* (Shook 2008, Sobek & Smalley 2008) is an effective way to ensure the widest level of collaboration in and commitment to decisions and to document the decision process and consequent learning so it can be used to create a library of research and ideas for current and future projects – a contribution to knowledge management.

The A3 process and Choosing-by-Advantages together support the idea that *decisions be made slowly with consensus so that whatever is decided can be implemented fast* (Liker's *principle 13* (2004)).

Evidence-based design

Supporting set-based design, evidence-based design (EBD) exists to help designers make a connection between design and the outcomes that owners want from their buildings. Data from an architect or owner with an axe to grind may not be the most reliable. EBD research seeks to establish causal relationships between design decisions and desired corporate outcomes. Still in its infancy, EBD is most fully developed in healthcare where evidence from clinicians is available and meta-analyses are possible.

Choosing to use EBD is a commitment to basing design (generating, evaluating, selecting from alternatives) on the best available evidence, and to actively search for and create that evidence. Hence it can be said to be a commitment to *research-based design*.

As Rybkowski reports (2009), the publication of Roger Ulrich's 1984 *Science* paper 'View through a window may influence recovery from surgery' marked the beginning of the EBD field. Ulrich analyzed the recovery records of forty-six surgical patients assigned to one of eight rooms. The recovery rooms were identical in all ways but one. On each floor, windows of half of the rooms faced a brick wall, while half faced a natural scene (Figure 7).

By systematically controlling for a range of other factors, the forty-six patients were matched into twenty-three pairs. Comparison of recovery rates indicated statistically significant differences; patients whose windows faced foliage had

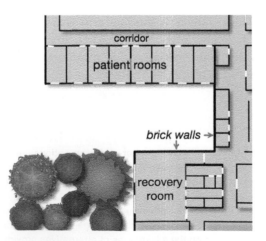

Figure 7 Floor plan of hospital showing patient rooms facing foliage & those facing brick wall. After Ulrich (1984), reprinted with permission from AAAS.

Figure 8 Three basic domains of project delivery. Graphic: Alan Mossman. Reproduced by permission of the authors.

- shorter postoperative stays,

- received fewer negative evaluative comments in nurses notes, and

- took fewer potent analgesics *than their matching counterparts.*

Ulrich's results showed that it was possible to measure health benefits and financial consequences of design decisions.

The challenge now is to extend EBD to other building types.

Gluing it all together

Figure 4 includes the key elements of a lean project delivery process – some will be seen in historic project delivery environments. What is distinctive about lean project delivery is the way all the elements work together

The three domains in Figure 8 are present in all projects. In Lean Project Delivery there is a concerted effort to get the three domains working together for the good of the project and the client. The elements that populate the three domains on a lean project are likely to be different from those used on historic ones – an operating system like Last Planner, relational rather than transactional commercial terms (see below) and a collaborative organisation with integrated, high performing teams and open and integrated governance.

What is important is to ensure that the elements in these three domains work well together both within and between domains.

In a 2005 paper, Owen Matthews and Greg Howell, a Florida based constructor, listed four problems with the 'traditional' contractual approach:

- ***Good ideas are held back** – late involvement of specialist constructors deprives the design team of the opportunity to develop innovations with those who will deliver the project.*

- ***Contracting limits cooperation and innovation** – . . . the system of subcontracts that link the trades and form the framework for the relationships on the project . . . detail exactly what each subcontractor [is] to provide . . ., rules for compensation, and sometimes useful, if unrealistic, information about when work [is] to be performed. The 20 to 30 page subcontracts mostly [deal] with remedies and penalties for noncompliance. These contracts [make] it difficult to innovate across trade boundaries even though the work itself [is] frequently inter- dependent. (It is hard to have a wholesome relationship . . . when you have a charge of dynamite around your neck and the other holds the detonator.) Of course, horse- trading always takes place . . ., but for 'equal' horses. Trading a small increase in effort by one contractor for a big reduction for another, a horse for a pony, [is] almost impossible.*

- ***Inability to coordinate*** – . . . *no formal effort to link the planning systems of the various subcontractors, or to form any mutual commitments or expectations amongst them. Project organizations looked like 20 or more rubber balls, representing subcontractors, all tethered to a single point by long elastic bands. When the connection point [is] jiggled, the balls jiggled in all random directions colliding with each other in unusual and unexpected ways.*

- ***The pressure for local optimization*** – *each subcontractor fights to optimize their performance because no one else will take care of him. The subcontract . . . and the inability to coordinate drives us to defend our turf at the expense of both the client and other subcontractors. Remember that everyone on the project other than the prime contractor is a subcontractor. These subcontractors frequently, in their life outside of the subcontract, may be generous, caring and professional. However, since right or wrong is defined by the subcontract, more often than not, they take on a very legalistic and litigious stance becoming an army where the rules of engagement are 'Every man for himself.'*

That is the context for what follows:

Commercial terms

As noted above, in the words of Darrington et al (2009) 'traditional construction projects are comprised of many two- party contracts that create a vertical chain of relationships that flow back to the owner, but do not interconnect project participants across contractual lines. As a result of this contract structure, each participant operates under commercial terms that provide economic incentive for it to maximize its own interests regardless of whether its actions would hurt other project players or benefit the project as a whole.'

Traditional contracts are transactional. Construction is effected through relationships that encompass a myriad of transactions which is why lean constructors prefer to work with relational agreements that recognise the reality of what needs to happen for successful project delivery. There are now a number of different relational-type agreements:

- Integrated Form of Agreement for Lean Project Delivery (IFoA) (USA)[10]

- AIA C191–2009 Standard Form Multi-Party Agreement for IPD (USA)[11]

- ConsensusDOCS300 (USA)[12]

- PPC2000 & PPC2000 International (UK)[13]

- Alliancing Agreements (Australia)[14]

Relational agreements create a collaborative system with shared responsibility for managing risk and shared pain/gain tied to the amount of value generated by the end product.

The Integrated Form of Agreement requires the use of lean methods and Last Planner. None of the others do, though there are some in Australia who are keen

[10] http://www.mhalaw.com/mha/newsroom/articles/ABA_IntegratedAgmt.pdf 17apr10 Available from Will Lichtig upon request. http://www.mhalaw.com/mha/attorneys/lichtig.htm 1may10

[11] http://www.aia.org/aiaucmp/groups/aia/documents/pdf/aiab081563.pdf 17apr10 list of distributors from: http://www.aia.org/groups/aia/documents/pdf/aias076340.pdf 17apr10

[12] http://consensusdocs.org/catalog/300-series/ 17apr10

[13] http://www.ppc2000.co.uk/buyppc.htm 17apr10

[14] there is no standard form Alliance agreement – work is under way in Australia to create one.

to do so. The Terminal 5 Agreement used for the construction of London Heathrow's Terminal 5 was another example (as yet unpublished) of a relational agreement. From 1999 BAA required the use of Last Planner on all its projects.

Encouraging collective sharing of risks and cost savings, relational agreements enable parties to treat projects as collective enterprises, optimizing the project as a whole and enabling the movement of money across traditional commercial boundaries so that it is possible to trade a horse for a pony – i.e. for one *trade partner*[15] to spend €50k so that another can save €200k . This fosters an entrepreneurial mindset aimed at creating project value and allowing all to share in the savings.

Insurance

One potential issue with relational agreements is insurance. If each party to a relational agreement is required to have its own insurance and there is a claim during design or construction, an insurance company could force parties to sue one another in order to trigger insurance coverage, threatening relationships.

At Heathrow Terminal 5 the general contractor (who was also the client, BAA) took out a single project insurance that covered all parties *and* then worked actively with them to manage risk.

A recent, unpublished Australian Alliance Agreement (there is no standard form yet) states a clear intent to procure professional indemnity, public liability and works insurance for the project as a whole. Other insurances are the responsibility of the several parties.

PPC2000 International places responsibility for insurance of the project and the site on one member of the partnering team on behalf of the whole, but still expects each party to have third party and professional indemnity insurance (s.19).

Article 7 of AIA C191–2009 refers to *'integrated insurance products. . . . . . . structured to provide adequate coverage at reasonable cost, striving **to avoid duplication in coverage or exposure gaps**.'* (our emphasis).

S.21 of ConcensusDocs300 expects each party to have their own insurance and S.32 of the IFoA (v 9, 2009) requires *'Architect and Architect's consultants [to] purchase and maintain insurance'* and in Exhibit 5 states *'Contractor and its Subcontractors shall procure and maintain insurance . . . .'* Will Lichtig, author of IFoA, reports that Sutter Health is pursuing insurance that will cover risks for professional errors and omissions and commercial general liability from an integrated provider. The intention is that the policy will be written so that the parties will not be required to sue one another in order to trigger coverage.

We would like to see further inter-disciplinary research in this important area to establish which insurance models and practices will most effectively and economically enable the parties to a project to optimize the project as-a-whole.

Operating system

Lean project delivery works when individuals make and keep commitments – it doesn't work without it. Trust and relationships develop on the basis of reliable

---

[15] Lean practitioners tend to talk about *constructors* and *trade partners* rather than contractors and sub- contractors as, within a relational contract, there are no sub-contractors – everyone is a party to the same relational agreement.

promises. The Last Planner® System[16] LPS (e.g. Ballard & Howell 2003b; Macomber Howell & Reed 2005), is a *commitment management system* and its principal metric is *PPC,* a measure of planning quality, which is the percentage of promises (to do work on or before a specified day) completed when promised. A study by Liu & Ballard in 2008 showed a significant correlation between PPC and productivity in US engineering construction and in the same year Gonzalez et al demonstrated the same relationship in house building in Chile.

LPS was designed to improve the planning process in project-based production and create a more reliable production schedule. It does this by recognising that it is only worth doing detailed planning for a short period (a week or a day) and the people most suited to do it are those who will do the work – these are the *last planners* that give the system its name. The planning that last planners do is done within the context of higher-level plans that they have contributed to. These layers of increasingly detailed schedules and involvement create a context in which it is possible for the last planners to make and keep promises.

There are five key collaborative conversations that together make up the Last Planner System (Mossman 2009). Each brings its own benefits. When all are working together they reinforce each other and the overall benefits are greater. The key *conversations* are:

- **Collaborative pull-scheduling** – creating and agreeing the production sequence (and compressing it if required)

- **MakeReady** – Making activities ready so that they can be done when we want to do them.

- **Collaborative pull-based Production Planning** – agreeing production activities for the next day or week and making promises about when they will be completed

- **Production Management** – monitoring production to help keep all activities on track

- **Measurement, learning and continual improvement** – learning about and improving the project, planning and production processes by studying reasons for late delivery and activities that went better than expected.

Last Planner works in construction and design. There are also Last Planner derivatives developed especially for design such as Responsibility-based Project Delivery™ and the Design Delivery System™[17]. The existence of these and other proprietary processes is evidence of practice leading academia.

With the Last Planner commitment management process as its kernel, the lean operating system uses a range of other systems in support. We have already mentioned the A3 process and Choosing by Advantages. Building Information Modelling BIM is vital as it enables a range of conversations around virtual prototypes (Virtual Design & Construction VDC) and exploration of the possibilities of off-site fabrication as well as eliminating many assembly issues before they get anywhere near site. Virtual *First Run Studies* (Nguyen 2009) and Virtual *Value Stream Mapping*[18] can help to establish safe and effective assembly sequences during design. We discuss *Target Value Design* below.

---

[16] Last Planner is a registered mark of the Lean Construction Institute www.leanconstruction.org

[17] Responsibility-based Project Delivery RbPD is a trademark of Lean Project Consulting www.leanproject.com; Design Delivery System DDS is a trademark of The Change Business Ltd www.thechangebsiness.co.uk

[18] for a discussion of Value Stream Mapping reality see Rother and Shook 1999.

Organisation

In addition to the commercial terms that encourage teamwork and an operating system that requires teamwork, an organisation that supports the formation of effective project- level teams is needed to enable the team members to become advocates of the project no matter who pays their salary.

Creating a unified project culture from individuals who come from – and at the end of the project will return to – a diverse range of organisational cultures is challenging. How can we create a team of project advocates from individuals who owe allegiance to many different organisations – a *superteam*[19]?

Factors assisting the creation of a unified project culture include:

- **Co-location** – bring all the key players together for the duration of a larger project and for continuous periods of 3-5 days at regular intervals on smaller ones.

- **Integrated governance** – Manage the whole team jointly – on the San Francisco Cathedral Hill Hospital project, a core team of five – the Sutter Health Program Manager, the Sutter Health affiliate (end-user), lead designer, lead constructor, and concrete constructor – provide project governance; at Heathrow T5 staff were given roles that befitted their skills irrespective of who paid their salary. One reason effective governance (leadership) is so vital is that roles and responsibilities are changing. This new way of working challenges existing boundaries and needs project team members with different skill sets to work in different ways. Significant and continuing investment of time and thought from the project leadership team is essential.

- **Align aspirations** – do things that simultaneously move the project forward in social or technical terms and build mutual understanding, respect and a shared vision and culture. The Study-action Team[20] approach is one way to do this (Hill et al 2007).

- **Practice supervision** – to develop people (e.g. mentoring and coaching), develop processes and enforce the use of proven processes. Particularly important is how supervisors behave in response to breakdowns—do they seek the guilty or seek the cause? (For deeper discussion of this topic see Mann 2010, 91-94; Ryan & Oestreich 1998; Dekker 2006; Abdelhamid et al 2009)

In all the recent projects discussed here owners/clients representatives are actively involved in project leadership. Sometimes, as in CHH, the eventual user is involved too. This link to the customer appears to us to be important.

What we see is that leadership (or project governance) is important for making integrated design and delivery work and, more particularly, that consistency of leadership is crucial (e.g. *leader standard work* Mann 2010, 37ff).

Commercial terms, operating system and organisation concern the soft aspects of construction – people and processes – the bits that are hardest to research, hardest to make money out of and hardest to get right. And this is not about fitting the people to the technology – its about designing the technology to work with the people.

---

[19] Colin Hastings *et al,* (1986) *Superteams: A Blueprint for Organisational Success* Fontana. For a summary see www.thechangebusiness.co.uk

[20] Study-action Team is a trademark of Lean Project Consulting www.leanproject.com

Appendix A

Table 3 Conditions to help create a high performing team on the Cathedral Hill Hospital Project as reported by Long et al 2007.

Creating a high performing team

- Commitment to deliver value
- Create an environment for learning
- Challenge paradigms
- Make decisions effectively
- Design for change free construction
- Embrace innovation
- Environment of trust, honesty and continuous improvement
- One owner voice
- Implement Lean Project Delivery and lean leadership
- Planning process that is based on *Network of Commitments*[21]

Table 4 Comparison of St. Olaf Fieldhouse and Carleton College Recreation Center. (after Ballard and Reiser 2004).

	St. Olaf College Fieldhouse	Carleton College Recreation Center
Contract type	Design-Build	Design-Bid-Build
Lean construction	yes	no
Last Planner	yes	no
Completion Date	August 2002	April 2000
Project Duration	14 months	24 months
Gross Square Feet	114,000	85,414
Total Cost (incl. A/E & CM fees)	$11.7m	$13,5m
Cost per square foot	$102.79	$158.44

The second Egan Report (2002) set a target for 20% of UK construction projects by value to be undertaken by integrated teams and supply chains by 2004, 50% by 2007. We believe that this didn't happen because the commercial terms, the operating system and the organisation were never aligned with this aim.

Target Value Design

As Darrington et al (2009) note Target Value Design is a collaborative strategy and process for designing based on the articulated project values, which become design criteria rather than mere aspirations. Within the TVD process, design is based on detailed estimates, rather than estimates waiting for a detailed design. This requires new skills – the ability – and willingness – to provide estimates on the basis of incomplete & conceptual designs.

> One of the earliest examples of TVD was St Olaf College Fieldhouse where a close approximation to 'redoing the same project with a different method' occurred when a different contractor built another fieldhouse with similar specification for a private college in the same city. As Ballard and Reiser (2004) note, the St. Olaf Fieldhouse delivery team integrated Lean Construction principles and practices including target costing and Last Planner. Table 4 shows the comparison.

The Target Value Design process has developed considerably since then.

Figure 9 shows the key stages of the TVD process. It is the primary methodology used to manage the definition and design phases of the Lean Project Delivery System.

---

[21] *Network of Commitments* is an integral part of the Last Planner commitment management system.

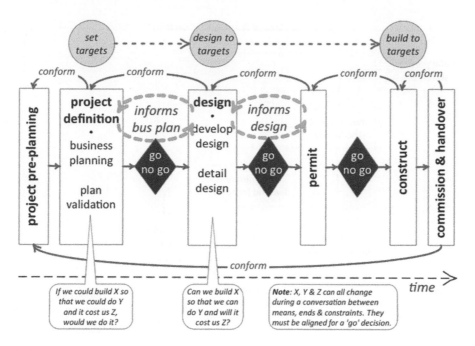

Figure 9 The Target Value Design Process (after Ballard 2009 with additional material from Long *et al* 2007. Graphic: Alan Mossman. Reproduced by permission of the authors.)

After initial project pre-planning by the owner, the TVD process starts with a project definition phase so now we will review the four phases of the Lean Project Delivery Process (Figure 4) in the context of *Target Value Design* and using the elements we have already introduced.

Project definition

The project definition phase seeks to establish a shared understanding of the business case for the proposed building or structure, an allowable cost and time, and to ensure that the project is *doable* within that cost and time. This process involves the client system in building a picture of the activities they envisage in the new facility – it may even involve the improvement of existing processes so that the new building is able to accommodate them from day 1.

Figure 10 is a flowchart of the project definition process. Notice that it starts with purpose—something that some clients seem unwilling to share with their project teams. Without understanding purpose, project team members will find it difficult to visualise the benefits, outcomes and interests ('D' in Figure 5) that the new facility will enable.

The client system will let the project team know when project definition is complete – it will be when they feel that the team has the necessary understanding of what is required. Once the project team are clear about the client system's interests and the need for the facility and how much the client is willing to pay for it – the *allowable cost.* Baseline expectations are explored for ends (what's to be delivered) and constraints (typically time and cost), and the team attempts to validate whether the ends can be provided within the constraints so that they can commit to the design and delivery.

The *target value* for a project is a statement of the interests that the client system wants a new facility to address. It is rare that client systems are able to articulate their interests from the outset – they often need help to do that in conversations

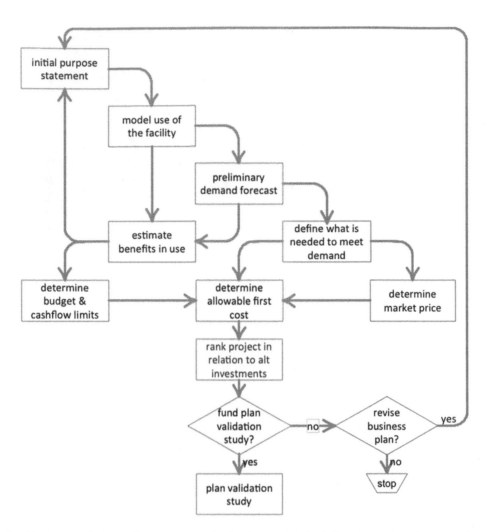

Figure 10 the Project Definition Process (after Ballard, unpublished. Graphic: Alan Mossman. Reproduced by permission of the authors.)

with the project team. As conversations progress interests may change. That is a reason to keep design options open to the *last responsible moment.*

If the client decides to fund a plan validation study, key members of the team that will deliver the project if funded (architects, engineers, general contractor, specialty contractors, suppliers) are engaged through professional services contracts to work with the client to improve and validate the business plan. The client says 'I want X within constraints Y.' The validation study, conducted by the team (including the owner, designers and constructors) that will build the project if approved, determines whether or not that is possible or advisable. What's wanted may change through consideration of new alternatives or through better understanding the consequences of what's desired. It is essential to align ends and constraints so design can create means to match. This is high level *value engineering*[22] (VE) done at the earliest stage in the project when it can have the greatest effect. VE continues throughout the project.

In the course of the project, whenever ends, means or constraints get out of alignment, they must be realigned or the project cannot be managed. That does not

---

[22] In the UK *value engineering* is known as *value management*

mean resisting change, but is rather a way of understanding how to manage change.

Lean design

Macomber *et al* (2008) suggested nine practices that together help create the conditions for a design process to deliver the *target value* within the client's *target price*:

- **Engage deeply with the client to establish the target value**.
- **Lead the design effort for learning and innovation**.
- **Design to a detailed estimate**.
- **Collaboratively plan and replan the project**[23].
- **Concurrently design the product and the process in design sets** (see above).
- **Design and detail in the sequence of the customer who will use it**.
- **Work in small and diverse groups to support learning and innovation**.
- **Co-locate design team members in a *big room***.
- **Review and reflect throughout the process**.

If the project business plan is validated and the owner goes forward with the project, the team works to create what's wanted at an *expected cost* less than or equal to the *allowable cost*. Some teams set themselves stretch goals to promote innovation, for example, setting target cost lower than allowable cost in order to fund scope & value additions.

> On the Sutter Fairfield Medical Office Building the user's representative had no prior experience with TVD so, according to Rybkowski (2009), it was not surprising that he challenged the process only three months in after receiving professional service invoices earlier than anticipated and the project estimated cost was higher than the allowable. In a tense meeting, Mike Tesmer, Director of Preconstruction Services, Boldt Company, explained that low early estimates on most design-bid-build projects tend to increase later as details are added to the design. By contrast, he argued, initially high estimates on the Fairfield project would probably drop as progressive *value engineering* trade-offs by the full professional team early in design allowed them to agreed design details and minimize contingency. To illustrate his point, Tesmer sketched a diagram (Figure 11) explaining how the two delivery systems differ.

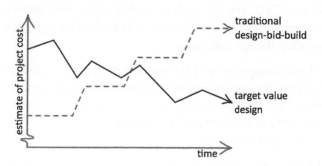

Figure 11 The Tesmer Diagram (after Rybkowski 2009, 140. Graphic: Alan Mossman. Reproduced by permission of the authors.)

---

[23] use of *Design Structure Matrices* can help avoid delay, rework and out-of-sequence design, Pull-scheduling will then help with the detail – see Last Planner in *operating system* above.

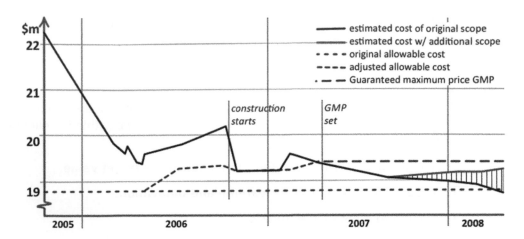

Figure 12 Project estimates over time for the Sutter Fairfield MOB Project [Project completed in 25 months, despite 3 month construction start delay. Target cost ($18.9m) was set 14% below market ($22.0m). Actual cost ($17.9m) for original scope was 5% below target and 19% below market. Additional scope ($.45m) represented by vertical hatching.] Graphic: Alan Mossman. Reproduced by permission of the authors.

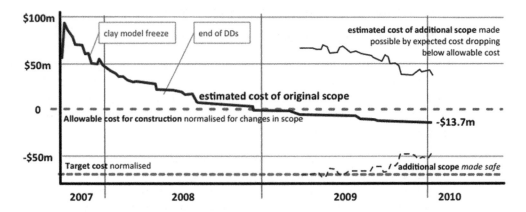

Figure 13 Project estimates to Jan 2010 for Sutter's $911m Cathedral Hil Hospital. Allowable cost is 13% below market. Construction has yet to start. (source: HerreroBoldt. Graphic: Alan Mossman. Reproduced by permission of the authors.)

Figure 12 shows the reality of what happened subsequently.

Design proceeds within the integrated team. There is a weekly schedule of formal meetings – TVD, weekly work planning, etc – and between these, ever-changing groups meet formally and informally to explore design problems of mutual interest and develop design solutions.

Some have suggested that the reduction in cost achieved so far on CHH is attributable to the recession. Figure 14 shows the US Army Corps of Engineers building cost index for the relevant period. In the time that the CHH team have reduced their estimate of out-turn cost to around 15% below market, the Corps of Engineers index has risen by 3%.

Constructors are integral to this free flowing but purposeful process. It is they who are best able to assess the prices for constructing the design and to negotiate alternative designs that might be built more cheaply while maintaining the designer's concept.

Figure 13 shows how, by using TVD with SBD & EBD, the estimated cost of the original scope changed over time – it is reviewed every two weeks – and how, when

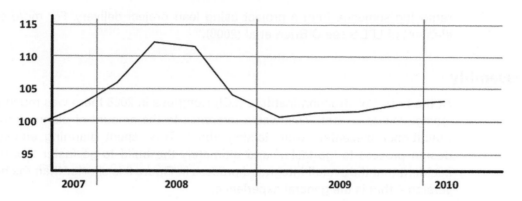

Figure 14 US Army Corps of Engineers cost index for buildings. 2007 Q3 = 100. Graphic: Alan Mossman. Reproduced by permission of the authors.

Appendix A

the estimate dropped below the target cost of construction, the project team began applying TVD, SBD to the list of items that were not part of the original scope but were on the client's *wish list* – and the expected cost of these items fell too. This is in complete contrast to what happens in *Design-bid-build* situations when bids come in above the allowable cost. What then tends to happen is that scope is cut and the client is later disappointed with the compromised facility as the expectations created in design are not met. TVD creates the possibility for client delight e.g.:

> Completed in 2006, the $13.1m Shawano Clinic is an outpatient facility with exceptional diagnostic capabilities, in part because of target costing. The project team delivered the project almost 15% below budget, itself 4% below market, and 3.5 months ahead of schedule, generating 70 additional days of clinic revenue equivalent to nearly $1m additional revenue for the owner and added service capability in imaging beyond the initial scope.

It is tempting to think that only constructors are at-risk when it comes to pricing – but that's not true. Constructors may be best placed to make and commit to prices – but everyone, even the designers, have an interest in helping them do it well as all share in the pain if constructors price too low and, if they price is too high, the client suffers – money left over that they could have used to expand the scope.

Another benefit from early constructor involvement is that the production system – how the work and its parts will be fabricated and installed – can be designed alongside the facility design and there is plenty of time to plan for off-site fabrication of elements and in the process reduce the number of personnel required on site (a safety bonus[24]) while potentially building faster with higher quality.

Lean supply

Because the production system is designed alongside the facility design, lean supply planning can begin then too. All the key players are already on board and involved in the design so they can plan logistics, prefabrication and the like.

The overall focus on making reliable promises is invaluable here. Real relationships develop around the project and it is difficult to renege on a promise in the context of a *real* emotional relationship. Unreliability is one of the few grounds for

---

[24] It is also an opportunity to use Safety by Design and Prevention by Design http://www.designforconstructionsafety.org 3may10

removing someone from a project using lean project delivery. For more on this element of LPDS see O'Brien et al (2008).

Lean assembly

At the Lean Construction Institute (LCI) Congress in 2008 there was much talk of *'going slow to go fast'*. This was a reference to the amount of planning that the practitioner presenters were talking about. Time spent planning enables the design process to 'go faster' and ultimately the build to go faster too. It is no accident that the overall integrated process in Figure 3 is shorter than the historic version – that is the general experience.

A high level of planning continues on site – all the Last Planner conversations listed above are just as important on-site as in the design office or *big-room*.

Because all involved are joined into the same contract, with shared risk and reward, the team is able to continue to drive the cost down as that will increase everyone's margin far more than local optimisation is likely to increase the margin of any one player.

Taken together all these elements help teams deliver ahead of schedule and below budgets that are already set below market. In traditional design and construction processes loaded with a cushion of surplus time, money and/or materials owned by the supply chain not the customer so customer has no chance to influence the use and distribution of that contingency

Table 5 shows the range of projects that have been or are being delivered using TVD. The list is dominated by hospitals for historical reasons rather than because the strategy is particularly suited to the design of medical facilities.

Integrated Project Delivery

Lean Project Delivery (LPD) is Integrated Project Delivery IPD[25]. IPD as interpreted by the American Institute of Architects (AIA) is not LPD. Organizational integration was included as an essential feature from the first LCI white paper on LPDS (Ballard 2000). Contracts and commercial terms were not specified. According to Cohen (2010), AIA's most recent definition describes IPD as *'a project delivery*

Table 5 examples of clients using TVD.

Owner/client	Project(s)
Childrens Hospital, Seattle, WA	Childrens' Hospital
Chinese Hospital, San Francisco	New Hospital
Sacred Sisters of Mercy, St Louis, MO	SSM Cardinal Glennon; SSM St. Clare
St Olaf's College, WI	Fieldhouse
Sutter Health for itself and on behalf of affiliates in Northern California	San Carlos Hospital; Fairfield MOB; ARC Roseville; – Use TVD on all current projects Cathedral Hill Hospital; Cathedral Hill MOB; St Luke's Hospital; Castro Valley Medical Centre;
ThedaCare, WI	Various
Universal Health Services	use some form of TVD on almost all projects; currently beginning a $140m full TVD project for completion early 2013
University of California San Francisco	Medical Center; Cardiovascular Research Center

[25] IPD a trademark of Westbrook Air Conditioning & Plumbing, Box 5459, Orlando, Fl 32855-5459. they have granted LCI the right to use it in trade.

Table 6 Cohen's criteria for project inclusion in his 2010 report.

Essential criteria	Desirable but not essential
• early involvement of key participants • shared risk and reward • multi-party contract • collaborative decision making and control • liability waivers among key participants • jointly developed & validated project goals	• mutual trust and respect among participants • collaborative innovation • intensified early planning • open communication within the project team • building information modeling (BIM) • lean principles of design, construction & operations • co-location of teams • transparent financials

Table 7 Comparison of IPD and historic project delivery approaches – features – after AIA 2007 & Vanguard 1999; from www.thechangebusiness.co.uk & www.leanconstructon.org.uk.

Integrated/lean project delivery		Historic project delivery
Learning, continual improvement, engaging with reality	CULTURE	Blame, finger pointing, exploiting loopholes, individual reward maximisation, risk averse
Systems thinking; Optimise the whole; encourage, foster & support multi-lateral open sharing & collaboration	THINKING	Command & control; encourage unilateral effort; Break project into constituent parts; Optimise parts (especially 'my bit')
Outside-in: act on the system to improve it *for customers* (helped by those working in it).	MANAGEMENT ETHOS	Top-down: Manage the contract, manage the programme, manage budgets, manage people
Integrated with work; based on data	DECISIONS	Separated from work
Related to purpose, capability & variation	MEASURES	Budget output, activity, standards, productivity
Based on demand, value & flow; open, collaborative & integrated team of key players formed at the outset & added to as the stakeholder group grows	ORGANISATION DESIGN	Functional specialisation; fragmented, silo based, strongly hierarchical, controlled; constructors not generally added until late in process
Concurrent & multi-level; high trust & respect	PROCESS	Linear, distinct, siloed (over-the-wall);
Shared openly & early	KNOWLEDGE & EXPERTISE RISK	Gathered 'just-as-needed', hoarded in silos
Collectively managed, appropriately shared	RISK	Individually managed, transferred as much as possible
Team success tied to project success; value-based	COMPENSATION & REWARD	Individually pursued; min effort for max return; (usually) first-cost based
Digitally based, virtual; Building Information Modelling (3, 4 & 5D); Short- term planning e.g. Last Planner	COMMUNICATION TECHNOLOGY	Paper-based, 2 dimensional; analog;
What *matters* to them? – Understanding their *human* & technical concerns.	ATTITUDE TO CUSTOMERS	Contractual

method distinguished by a contractual agreement between a minimum of the owner, design professional, and builder where risk and reward are shared and stakeholder success is dependent on project success.'

Cohen's report *Integrated Project Delivery: Case Studies* describes six cases, three of which actively aligned lean project delivery, commercial terms, operating system and organisation. Table 6 summarises Cohen's criteria for project inclusion.

Table 7 is a further summary of the differences between Lean IPD & historic DBB delivery.

Issues for further research

In research we would like to see:

- studies drawing on the TVD experience in Finland and UK as well as the US

- studies to extend the idea of TVD to embrace whole-of-life value – this might involve a variable allowable cost so that project teams can negotiate tradeoffs between first cost and the costs and benefits of the use of the facility.

- more *Action Research* approaches[26] that will help us to understand *'what cannot be seen'* (Pavez & Alarcon 2008) and enable us to be much more aware of the needs of the people in our industry.

- An understanding of who benefits from waste in the end-to-end design & construction process.

- Exploration of the way that technological advances can be adapted to people so as to help all involved in creating our built environment to deliver safer, greener, better, faster and cheaper design and construction.

- Studies to broaden the scope for Evidence-based Design EBD

- Socio-technical studies of the application of Set-based design

- Social and ethnographic studies of BIM & VDC and the leadership required to make it work

- Inter-disciplinary studies of insurance models that support IPD

- Studies of the impact on site-based production of designing product and production system together

- Studies to explore the applicability of these ideas outside the US.

In the blog quoted earlier, Deutsch (2010) suggested 'BIM and Integrated Design offer an opportunity for the profession and industry to transform itself in ways unseen in over a century.' As Deutch recognises, this will affect the way those entering the profession are taught. Is it, he asks, still reasonable to assume *'that students pick [BIM] up as though it were any 3D software',* and that they can, as if by osmosis, understand *'how to work effectively with others in a BIM environment?'*

In Academic environments we would like to see disciplines learning together for at least some of the time so that they understand and appreciate each other's skills, knowledge and jargon.

We have technology, we have processes, commercial terms, operating system and organisation (all of which can be improved)— what we now need is more research to establish whether or not this is already a better way to build in a range of settings and for a wider range of project types – and, if it is better, how it can be improved.

In his 1962 report Sir Harold Emmerson reminded his readers that a 1950 UK Building Working Party – *The Allen Report* –recommended that those entering the building industry should take a common course of study for an initial period. Perhaps the time has now come for *integrated education* for built environment professionals.

---

[26] http://en.wikipedia.org/wiki/Action_research describes a range of sources

Conclusions

We have described *learning* in practice around a number of collaborative processes that enable integrated design and delivery of projects in the built environment in the US.

We believe that these are robust processes that could be used in other contexts and with certain other building types.

These inter-related and collaborative processes are integral to Lean Project Delivery. A number of these processes have been developed by or with practitioners and it is practitioners who are developing the skills and the leadership to get them working together.

Strong leadership that involves members of the client system appears to be critical for making this work.

Two outcomes of TVD look to be repeatable (at least within healthcare and education):

1. Projects are completed below market cost—so far as much as 19% below.

2. Estimated out-turn cost falls as design develops.

Acknowledgments

Thanks to Ade Tomlinson, Project Manager, Juliet Odgers, Cardiff School of Architecture, Daria Zimina, Loughborough University, Jose Jorge Ochoa Paniagua, Hong Kong Polytechnic University, Will Lichtig, McDonough Holland & Allen; Prof Tariq Abdelhamid, Michigan State University; Professor Robert Amor, The University of Auckland for invaluable comments on earlier drafts.

References

Note: All papers from International Group for Lean Construction (IGLC) conferences since 1996 are available at http://www.iglc.net

AIA (2007) *Integrated Project Delivery: A guide* AIA http://www.aia.org/contractdocs/ AIAS077630 9apr10.

Abdelhamid, Tariq, Don Schafer, Tim Mrozowski, V Jayaraman, Greg Howell & Mohamed A.El-Gafy (2009) 'Working through unforeseen uncertainties using the OODA loop: an approach for self-managed construction teams.' *Proceedings of IGLC-17*, Taipei, Taiwan.

Alexander, Christopher (1974) *Notes on the Synthesis of Form 2e.* Harvard MA.

Ballard, Glenn (2000) *LCI White Paper 8: Lean Project Delivery System* Lean Construction Institute. http://www.leanconstruction.org/lpds.htm 5Apr10.

Ballard, Glenn (2009). 'TVD Update', a presentation at the June, 2009 Lean Design Forum, St. Louis, MO. Available at www.leanconstructionorg/files

Ballard, Glenn & Greg Howell (2003a) Lean project management *Building Research &Information* (2003) 31(2), 119–133.

Ballard, Glenn & Greg Howell (2003b) An update on Last Planner *Proceedings IGLC11, Blacksburg, VA.*

Ballard, Glenn, Iris Tommelein, Lauri Koskela & Greg Howell (2002) '*Lean Construction tools & techniques*' chapter 15 in Best & Valence eds (2002) p227ff.

Appendix A

Ballard, Glenn & Paul Reiser (2004) The St. Olaf College Fieldhouse Project: A Case Study in designing to target cost in *proceedings of IGLC-14*, Helsingor, Denmark.

Best, Rick & Gerard de Valence eds (2002) *Design & Construction: Building in Value* Butterworth-Heinemann Elsevier Oxford UK.

Cohen, Johnathan (2010) *Integrated Project Delivery: Case Studies* The American Institute of Architects, California Council in partnership with AIA http://www.aia.org/about/initiatives/AIAB082049 9apr10.

CURT (2004) *Collaboration, Integrated Information, and the Project Lifecycle in Building Design and Construction and Operation Construction* Users Roundtable, USA WP-1202.

Darrington, Joel, Dennis Dunne & Will Lichtig (2009) *Organization, Operating System And Commercial Terms* in Thomsen, Chuck, Joel Darrington, Dennis Dunne & Will Lichtig (2009) *Managing Integrated Project Delivery* Construction Managers Association of America, Mclean, VA.

Dekker, Sidney (2006) *The field guide to understanding human error*. Ashgate UK

Deutsch, Randy (2010) Notes on the Synthesis of BIM *AECbytes Viewpoint #51* (April 7,2010) http://www.aecbytes.com/viewpoint/2010/issue_51.html

Eckblad, Stuart, Jim Bedrick, Zigmund Rubel (2007) *The Possibilities of an Integrated approach* Presentation to the AIA California Council Change Conference, June 25–26, San Francisco.

Economist (2010) The art of concealment Mar 18th 2010 www.economist.com/world/britain/displaystory.cfm?story_id=15731336 5Apr10.

Egan, Sir John (2002) Accelerating Change A report by the Strategic Forum for Construction, London. www.strategicforum.org.uk/pdf/report_sept02.pdf 8Apr10

Emmerson, Sir Harold (1962) *Survey of problems before the construction industry*: The Emmerson Report HMSO for the Ministry of Works.

Evans, R., Haryott, R., Haste, N. and Jones, A. (1998) *The Long Term Costs of Owning and Using Buildings*, Royal Academy of Engineering, London.

González, Vicente, Luis Fernando Alarcón & Fernando Mundaca (2008) 'Investigating the relationship between planning reliability and project performance', *Production Planning & Control* 19(5) 461–74.

Henderson, Bruce (1974) 'The Experience Curve Reviewed: V. Price Stability'. *Perspectives*. The Boston Consulting Group. http://www.bcg.com/impact_expertise/publications/files/Experience_Curve_V_Pric e_Stability_1973.pdf. 9apr2010

Hill, Kristin, Christine Slivon & John Draper (2007) Another approach to transforming project delivery: Creating a shared mind *Proceedings IGLC-15*, Michigan, USA.

Hughes, W., Ancell, D., Gruneberg, S. and Hirst, L. (2004) *Exposing the myth of the 1:5:200 ratio relating initial cost, maintenance and staffing costs of office buildings*. Paper presented at the ARCOM conference.

Ive, Graham (2006) Re-examining the costs and value ratios of owning and occupying buildings *Building Research & Information* 34(3), 230–245.

Kennedy, M.N. (2004) *Product Development for the Lean Enterprise*. Oaklea Press, Richmond, Virg., 256 pp.

Lichtig, Will (2007) *Creating a Relational Contract to Support Lean Project Delivery* presentation to the Lean Construction Institute Relational Contracting Meeting Chicago, IL June 14–15.

Liker, Jeffrey K (2004) *The Toyota Way: 14 Management Principles from the World's Greatest Manufacturer*. McGraw-Hill.

Long, David, Stephen Pepler & Paul Reiser (2007) *Cathedral Hill Hospital Validation Phase Review* presentation to the Lean Construction Institute Annual Congress http://www.leanconstruction.org/files/LCI_Seminars_and_Annual_Conferences/ 9thA nnualCongress/1_Plan_Validation/1_Plan_Validation_Cathedral_Hill.pdf 10apr10

Macomber, Hal, Greg Howell & Dean Reed (2005) Managing promises with the last planner system: closing in on uninterrupted flow. *Proceedings IGLC-13*, Sydney, Australia.

Macomber, Hal, Greg Howell & John Barberio (2008). *Target-Value Design: Nine Foundational Practices for Delivering Surprising Client Value.* http://www.aia.org/ nwsltr_pm.cfm?pagename=pm_a_112007_targetvaluedesign 6apr10

Mann, David (2010) *Creating a Lean Culture: tools to sustain lean conversations.* 2edn. CRC Press

Matthews, Owen & Greg Howell (2005) *Integrated project delivery An example of relational contracting.* Lean Construction Journal Vol 2 #1 April 2005 www. leanconstructionjournal.org 1may10.

Mauck, Robert, William A Lichtig, Digby R Christian & Joel Darrington (2009) Integrated Project Delivery: Different Outcomes, Different Rules paper presented to *The 48th Annual Meeting of Invited Attorneys* www.schinnerer.com/risk- mgmt/ Documents/AMIADocuments/2008-48thProceedings/Mauck-Lichtig-09-ipd.pdf 31mar10

Morgan, JM, & Jeffrey K Liker (2006) *The Toyota Product Development System: Integrating People, Process, and Technology*, Prod. Press, New York, NY, 377 pp.

Mossman, Alan (2009) *Last Planner: collaborative conversations for reliable design and construction delivery* http://www.thechangebusiness.co.uk/sites/default/files/ downloads/LastPlanner2009-05-21.pdf 14Apr10.

Nguyen, Hung V, Baris Lostuvali & Iris D. Tommelein (2009) Decision analysis using virtual first-run study of a viscous damping wall system. *Proceedings IGLC17* Taipei, Taiwan.

O'Brien, William J, Kerry A. London, Carlos T. Formoso, Ruben Vrijhoef – eds (2008) *Construction Supply Chain Management Handbook.*? CRC Press.

Owen, Robert ed (2009) *Integrated Design and Delivery Solutions*. CIB White Paper on IDDS, CIB Publication 328.

Parrish, Kristen and Iris Tommelein (2009) Making design decisions using Choosing by Advantages *Proceedings IGLC17* Taipei, Taiwan

Pavez, Ignacio & Luis F. Alarcón (2008) Lean Construction Professional's Profile (LCPP): Understanding the competences of a lean construction professional *Proceedings IGLC-15, Michigan, USA.*

Rother, Mike & John Shook (2009) *Learning to See: Value stream mapping to add value and eliminate muda*. Lean Enterprise Institute, Brookline MA.

Appendix A

Ryan, Kathleen D & Daniel K Oestreich (1998) Driving Fear out of the workplace: Creating the high-trust, high performance organisation. 2edn. Josey Bass, San Francisco.

Rybkowski, Zofia (2009) *The Application of Root Cause Analysis and Target Value Design to Evidence-Based Design in the Capital Planning of Healthcare Facilities* PhD Thesis University of California, Berkeley.

Salvatierra-Garrido Jose, Christine Pasquire & Tony Thorpe (2009) 'Value in construction from a lean thinking perspective: current state and future development' in *proceedings IGLC-17*, Taipei, Taiwan, www.iglc.net 5Apr10.

Saxon, Richard (2005). *Be Valuable: A guide to creating value in the built environment*. Constructing Excellence in the Built Environment, London. 50 p.

Shook, John (2008) *Managing to Learn: Using the A3 Management Process to Solve Problems, Gain Agreement, Mentor, & Lead*, Lean Enterprise Instit., Cambridge, MA.

Sobek, Durward K & Art Smalley (2008) *Understanding A3 Thinking: A Critical Component of Toyota's PDCA Management System*, CRC Press, New York, NY.

Sobek, Durward K, Allen C Ward & Jeffrey K Liker (1999) 'Toyota's Principles of Set-Based Concurrent Engineering.' *Sloan Management Review*, 40 (2) 67–83.

Suhr, J. (1999) *The Choosing By Advantages Decisionmaking System*, Quorum, Westport, CN.

Toyota Motor Corporation (2001) *The Toyota Way 2001*. Unpublished internal document, Toyota.

Ulrich, R. (1984). 'View through a window may influence recovery from surgery.' *Science*, 224(4647), 420–421.

Vanguard (1999) *The Vanguard Guide to Understanding your organisation as a system* Privately published, Vanguard Education, Buckingham UK.

Ward, Allen C (2007) *Lean Product and Process Development*, Lean Enterprise Institute, Cambridge, MA.

Ward, Allen C, Jeffrey K Liker, John J Cristiano & Durward K Sobek II (1995) 'The Second Toyota Paradox: How Delaying Decisions Can Make Better Cars Faster'. *Sloan Management Review*, Spring 1995, pp. 43–61.

Appendix A

APPENDIX B The 'ADePT' methodology for planning and managing the design stage of projects – Paul Waskett and Andrew Newton*

Introduction

For over 15 years the separation of design from the rest of the project process has been recognised as a fundamental weakness in the construction industry. Since then, steps have been taken by the industry to better integrate the design stages of projects within the broader project. This has, in part, been driven by wider involvement of contractors in the design process, but also by a growing awareness of the complex interaction between design and construction. However, the latest figures published by Constructing Excellence and the Department for Business, Innovation & Skills (1) show that the design still goes over budget in 20% of projects and is delivered late over 40% of the time, with the knock-on effect that only 63% of construction processes are delivered within budget and only 45% are on time. Clearly there is more to do.

The ADePT methodology

The 'ADePT' methodology is a highly structured approach to planning and managing the design activities within a project. It was developed in the 1990s and early 2000s through collaboration between industry and academia, including AMEC, Sheppard Robson, Arup, Laing and Loughborough University. The methodology has been exploited in the construction industry since 2002 by Adept Management Ltd, which was set up for that purpose. The ADePT methodology is built on three key premises:

- Design is often a largely repeatable process and so, using learning across projects, meaningful plans can be put in place;

- The development of design is driven by the production and exchange of information and so it is this information flow that should be planned and managed; and

- Design decisions and activities can be interdependent, leading to necessary iteration within the design process, and this iteration must be planned and managed.

The methodology comprises four stages:

Define:
Processes &
dependencies

Streamline:
Optimise decision-
making process

Plan:
Project & discipline
schedules

Deliver:
Management &
reporting

*Directors, Adept Management Ltd. www.adeptmanagement.com

The Design Manager's Handbook, First Edition. John Eynon.
© 2013 The Chartered Institute of Building. Published 2013 by Blackwell Publishing Ltd.

Stage 1: the scope of the design process and dependencies between activities are defined.

Stage 2: the sequence of the process is determined based on the dependencies between activities and the iteration within the process.

Stage 3: the design process is represented in the form of a schedule, enabling the integration of the design process with procurement and construction.

Stage 4: the design process and the flow of work are controlled.

These stages are described below in a little more detail.

Stage 1 – Defining the Scope of the Design Process

Clearly to define the full scope of a design project in terms of the activities required and the dependencies between them can be a time-consuming process. However, it has been possible to define the design process generically for most project types, with the majority of a project's activities predefined. In fact, the term 'generic' is somewhat misleading in that we hold a data-library that is being expanded continually as more projects are planned and, consequently, more parts of the process (which had previously been undefined) are added.

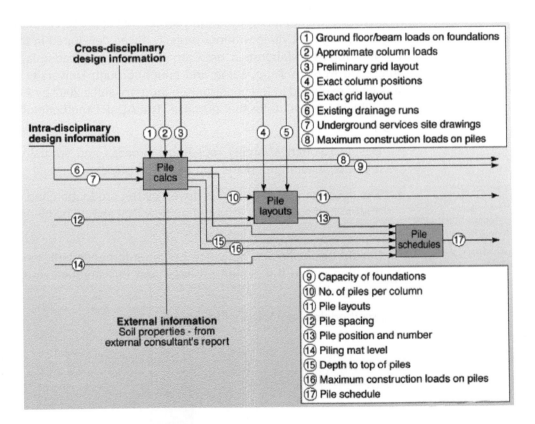

Stage 2 – Process Sequencing

A sequence of activities is calculated that minimises the iteration in the design process and ensures any assumptions that the team need to make are ones that can be made with confidence. This is achieved by weighting the dependencies between activities. The calculation of a sequence, including clusters of interrelated tasks, prioritises the availability of outputs associated with the most critical dependencies.

The interdependent, iterative groups of activities that remain in the process following sequencing are typically multi-disciplinary. They represent points in the design process where design team members should work concurrently to solve the interdependent problem. Usually they also represent elements of the construction, and therefore of the design output, which require co-ordination.

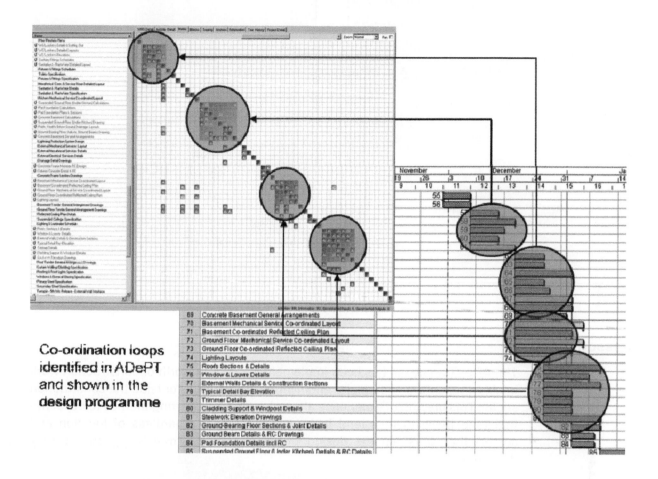

Co-ordination loops identified in ADePT and shown in the design programme

Stage 3 – Scheduling

The sequenced design process must be represented in a schedule so that design delivery dates can be seen alongside construction and procurement target dates. The optimised process is usually imported into the project-management tool that the project's Planner is using to maintain the construction schedule.

Stage 4 – Controlling the Design Workflow

All too often deviation from the agreed design schedule means that very quickly it does not really represent the design process being undertaken. It is then close to impossible to implement action to get the process back to the target schedule and the deviation increases to the point where the schedule is meaningless. Our experience is that many conventionally planned projects suffer from this problem, which contributes to the general lack of confidence in design schedules. Therefore, having produced a target design schedule, the design process needs to be controlled. ADePT incorporates an approach to process control, based on Lean Production methods, which tracks information production and exchange and which pre-empts deviation from the target schedule by analysing constraints, which then allows the schedule to be kept up to date and used in a meaningful way, with potential risks to the project being mitigated in advance.

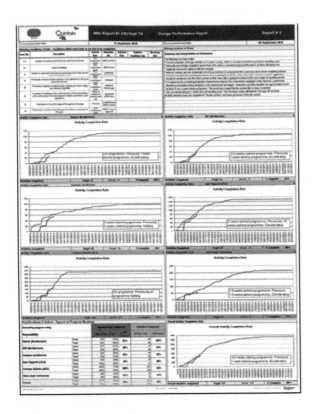
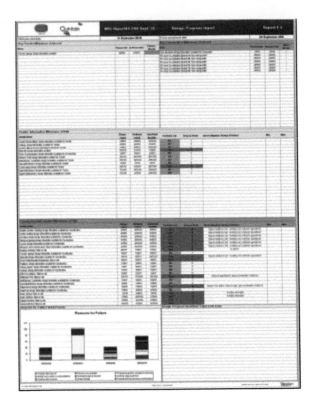

Lessons learned

The ADePT approach has been implemented on £5Bn worth of construction work in six continents, and on major engineering programmes outside the construction sector. This breadth of experience allows trends and recurring themes to be identified (and those themes tend to be broadly similar regardless of location and sector) that constrain the design phase of projects from being successfully delivered.

Those issues that we see most consistently include:

Client requirements / expectations / process

A lack of clarity over client requirements and expectations is a common theme. Without doubt the problem is amplified by the complex decision-making arrangements and number of stakeholders involved in many client bodies. Changes to requirements, or lack of certainty within the design team about those requirements, cause significant delays and rework.

Design team responsibilities

All too often we experience design-team members working to an ambiguous scope of responsibilities. Most frequently this concerns co-ordination requirements where multiple parties are required to have an involvement (such as design of rainwater systems), and also the level of detail which design consultants are required to produce.

Leadership

Many projects suffer from a shortage of clear leadership. As contractors take more contractual responsibility for the design stages of projects, we often see confusion over who takes responsibility for co-ordinating the design outputs and for leading the design team.

Design–Procurement Interface

There are a range of recurring problems created at the interface between the design process and procurement or commercial activities, including:

- Construction work packages are poorly defined at the time that design is planned, including acknowledgement of which packages are to involve a sub-contractor during the design stage;

- The level of information required to tender work packages is poorly defined;

- Designers are given insufficient information about the cost plan;

- Commercial decisions that affect design, such as which supplier is likely to be appointed, are not made in a timely manner; and

- Placement of orders is delayed in pursuit of the best commercial deal, meaning that subcontractor design, and the co-ordination of that design, are also delayed.

Fixing design/recording assumptions

We often see projects suffering from rework resulting from decisions (such as to fix an element of the design) or assumptions being revisited. A distinction needs to be drawn between that rework that benefits the project by resolving problems or refining an undesirable design solution, and rework that occurs without any real understanding of the benefit. Unfortunately this often undoes decisions or assumptions that were made previously and results in unexpected and unbudgeted rework.

Undefined overspend

Of course, many projects overspend for very many reasons and we, like others, have seen plenty of them. An issue we frequently come across is the problem of projects overspending by an unknown amount. This is likely to be a culmination of many other challenges combined with poor progress measurement, which means that the relationship between work done (and the resulting outputs) and effort expended or money spent is far from linear.

Future developments

The increased uptake of BIM

Over the years that the ADePT methodology has been in use, the way in which design is undertaken has changed. For instance, development of a whole-project thermal model, rather than individual calculations, is now the norm. We expect the whole design process to significantly change over the next few years as Building Information Modelling (BIM) becomes widespread. In fact, the design process should be even more recognisable as an information (or data) production process, the direct result of which is a populated BIM database. This will make attempts to plan and manage anything other than information even less meaningful than it is now.

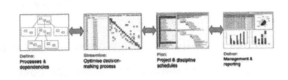

Information Planning: timely provision of design information

BIM: design information co-ordinated to the required level

The role of the Design Manager

We expect that the role of the Design Manager will shift towards that of 'model manager' or 'information manager', while still retaining aspects of design-quality management and design-team leadership. We also expect that the process of design development with BIM will be largely repeatable, as it is now, and that for design management to be effective the process will need to be meaningfully planned and skilfully managed.

The use of Earned Value Management in Design

We also anticipate a desire to make increased usage of Earned Value Management within the design stage of projects. To employ EVM in design in its true sense, a number of challenges must be overcome. For example, the design activities should be planned *and costed* in detail. Perhaps the largest challenge will remain the difficulty of measuring the Earned Value in design which, like all EV measurements, requires certainty that the outputs from activities have been produced to the required quality. When the output is information, which is often intangible, this assessment can be difficult to make. We expect more sophisticated ways of measuring design information to be developed than simply assessing progress against activities or the availability of deliverables. In fact, as information becomes more easily auditable in a BIM database, the uptake of BIM might enable truly effective implementation of EVM in design.

Reference

1. 2011 UK Industry Performance Report: Constructing Excellence www.constructingexcellence.org.uk/zones/kpizone/industryreport.jsp

Appendix B

APPENDIX C   Choosing by Advantages

Alan Mossman. Reproduced by permission of Alan Mossman, The Change Business Ltd.

Good decisions matter because they lead to effective actions that produce desired results. That's why the method you use to make decisions matters.

sound   *sound*   *effective*   *desired*
methods → **decisions** → **actions** → **outcomes**

Choosing By Advantages (CBA) enables decision makers to concentrate on what is important: the advantages (beneficial differences) that each alternative could deliver to stakeholders, and basing the decision on the total importance of those advantages. By focusing on advantages for the customer/end user of the project, CBA connects decision makers with their clients' ideas about what they want. Involving constructors ensures that 'buildability' is considered.

Developed by civil engineer Jim Suhr while employed by the US Forest Service, CBA is a system for making decisions that enables organisations, project teams and individuals to make more effective choices.

Why use CBA?

CBA creates an open, transparent and auditable decision process for design and construction work that acknowledges the complexity of most projects and of the client systems that commission them. CBA is well able to handle both objective and subjective data within a single decision process.

Design and construction projects are increasingly complex, rapid and uncertain. Client systems are increasingly complex too. No single person is 'the client'; rather, *the client* is made up of a complex and changing group of individuals with different, and probably changing, needs and expectations over the gestation of a project. A clear audit trail for decisions allows decisions to be revisited when necessary. A clear delivery programme allows all to understand the implications of changing those decisions.

As Sheena Iyengar has shown in *The Art of Choosing* (Twelve, 2010), humans have a tendency to shy away from too much choice. In design there is a tendency to leap to conclusions before all reasonable options have been explored – a strategy for avoiding *choice overload*. CBA offers a systematic way for all stakeholders to manage the process of deciding between a large number of alternatives without being overwhelmed, and ensures that we do not artificially limit the number of alternatives considered in the complex decisions we face in design for the built environment.

Using A3s to document your decisions adds further to the transparency of the decision process (if you have not come across an A3 before, it is a process of recording critical information, improvement processes, decisions, proposals, on one side of a single sheet of A3 paper. See John Shook's *Managing to Learn* (LEI, 2008). A3s are being used with increasing frequency in US design and construction organisations to ensure that design decisions are systematically arrived at and documented.)

What is CBA?

CBA is a system for basing decisions on the importance of *beneficial* differences (i.e. advantages) between alternatives. It has key definitions, models, principles and a set of methods. The key principles are:

1. Decisions must be based on the importance of the beneficial differences between alternatives.

2. Decisions must be anchored to relevant facts.

3. Different types of decision call for different *sound* methods of decision making

4. Decision makers must learn and skilfully use sound methods of decision making.

As principle 3 implies, there are a number of different methods for different types of decision. These range from simple binary decisions with no resource implications to complex ones with many alternatives, each with its own set of resource implications.

It is important to consider resource requirements in a different way from other attributes, as there is an important question for most stakeholders about what they would do with the resource if it wasn't consumed on the decision currently being considered. In CBA the beneficial differences of non-resource attributes of the alternatives are evaluated before any consideration is given to resources, so that any trade-offs can be clearly explored.

CBA avoids the pitfalls of unsound methods such as Kepner–Tregoe, choosing by pros and cons, using advantages *and* disadvantages, pair-wise comparison and weighting rating and calculating (WRC) systems, including criteria weighting, factor weighting and cost-benefit analysis.

What sort of situations call for CBA?

*   Deciding whether or not to bid or accept a contract

*   Selecting and managing projects and programs

*   Selecting consultants, contractors and suppliers

*   Selecting and purchasing materials, equipment, and other products

*   Choosing combination of design alternatives

*   Choosing between competing alignments for road and rail projects.

These are just some of the bigger choices. CBA can be applied to any decision. A good way to learn it is to use it for even the simplest instantaneous mentally-formed decisions. CBA basics are being taught in primary and secondary schools in the US.

In design and construction, construction companies like Boldt Construction, DPR, Herrero Contractors; designers like HKS, Smith Group, Boulder Associates (architects), CH2M Hill (engineers), and clients like Michigan State University, Sutter Health and UHS, the largest US healthcare provider, are using CBA to make sound decisions and A3s to document them. Examples include:

*   Universal Health Services – decisions about a number of capital projects

*   Sutter Health San Francisco Hospital designs – selection of seismic damping system

- Selecting trade partners/subcontractors

- Selecting combinations of structure, façade, vertical circulation, etc. in a *set-based design*

- Selecting policy and spending priorities.

The US Parks Service, which oversees many major built-environment decisions in sensitive contexts, describes CBA as 'a system of concepts and methods to structure decision making. CBA quantifies the relative importance of non-monetary advantages or benefits for a set of alternatives and allows subsequent benefit and cost consideration during decision making. CBA may be used as an evaluation method during the evaluation phase of the value analysis job plan, in lieu of the more traditional weighted-factor analysis. CBA is the preferred evaluation method where critical non-monetary benefits need to be evaluated.'

CBA and Value Management

Some in Value Management (it's called *Value Engineering* in the USA) have used CBA. CBA supports value optimisation and constructability decisions throughout the design process. It can also be used to find and evaluate new alternatives if bids come in too high in a conventionally tendered project.

SAVE, www.value-eng.org, recognises CBA and includes references to it. Only some of its practitioners follow it, as CBA demonstrates the unsoundness of some older methods such as those involving the weighting of factors. In the UK some Institute of Value Management practitioners use CBA though it is not one of the techniques mentioned on the IVM website, which features a number of unsound techniques.

How do I use CBA?

Simple processes are easy to learn and, once you have built the habit, easy to use.

For more complex decisions of the sort we find in design and construction, facilitation is useful and some training and mentoring for all participants is important in order that the meeting does not get bogged down with discussions about process that are covered in the training. John Koga, Director, Process and Supply Chain Management at HerreroBoldt, a construction joint venture, commented that 'Very few use CBA correctly without mentoring. They slip into incorrect and dangerous habits that are no longer CBA.'

Larger companies in the US are developing an internal facilitation capability to enable project teams to focus on content, while the facilitator holds the process as the team works through the five stages of the complete process[1]:

---

[1] Figure reproduced by permission of Alan Mossman, The Change Business Ltd.

Appendix C

I. Stage setting – defining the purpose and identifying the issues, the criteria of the decision and who should be involved

II. Innovation – identifying the alternatives and making the differences between them visible and tangible

III. Decision making – listing the advantages of each alternative, deciding the importance of each advantage and choosing the alternative with the greatest importance of advantages before considering the resource implications of the alternatives and making the draft decision.

IV. Reconsideration – reviewing the draft decision to check that it really is what is wanted, changing it if appropriate and then committing to the choice.

V. Implementation – doing what is necessary to realise the decision in reality.

The simplest decision processes may focus on stage 3 only, but most decisions will involve elements of all five stages.

Much work in stages 1 and 2 generally happens before the main meeting and is reviewed at the start of the decision-making stage. Stages 4 and 5 happen afterwards.

How can I learn CBA?

First, we teach people how to use correct data.
Second, we teach them how to use data correctly.

– Jim Suhr

Jim Suhr has developed a CBA training programme. There are a number of trainers with design and construction experience in the US and at least one in Europe. Suhr suggests that the learning process involves:

- ***Learning*** *just one set of CBA definitions, principles, models, and methods at a time.*

- ***Unlearning*** *(learning to not use) the corresponding unsound concepts and methods that the CBA concepts and methods are replacing.*

- ***Relearning*** *the CBA concepts and methods.*

- ***Practising and consistently using*** *the CBA concepts and methods that you have learned. (If possible, practise under the guidance of either a CBA facilitator or instructor.)*

- ***Teaching*** *the CBA concepts and methods to other people. This will not only be beneficial to them; it will also be beneficial to you. Then, return to step one and learn more.*

For more information, read Jim Suhr's *Choosing By Advantages* (Greenwood, 1999), download introductory papers from www.decisioninnovations.com/ or read one of Suhr's new introductory booklets; you can contact Jim via www.decisioninnovations.com.

APPENDIX D   Delivering value – A Guidance Note for Design Managers

Michael Graham

This paper aims to help Design Managers maximise value by introducing methods and strategies that deliver results and giving some practical tips. Project Examples and Signposts are also given for further consideration.

Every organisation manages and delivers value for its customers, employees, owners, and community. Every manager makes decisions that commit resources and (sometimes) add value.

The Design Manager is at the sharp end of decision making. Huge leverage is created by effective design management decisions, particularly at early stages in project development. And throughout project design and delivery the Design Manager will be conscious that each decision locks in cost, manages risk, and aims to deliver value.

This guidance note aims to help the Design Manager address the challenge to maximise value from available resources.

But first, *what is Value?* . . .

All managers are responsible for value. Organisations that set standards for practice achieve results. Value is not managed by chance.

Value is a measure of how well an organisation, project or product satisfies stakeholders' objectives in relation to the resources consumed. (EN – 1325-1[1])

Perceptions of value depend on the situation, vary between people and change over time. Different choices are made in emergency situations, where speed of response is vital, compared with situations where expenditure is planned in advance and a combination of quality, cost and customer satisfaction might control the choice made. The whole issue of stakeholder need and resolution of conflicting requirements must also be addressed.

Relatively small costs in design and construction can have huge implications for future users, whether expressed in financial, social or environmental terms.

The strategic choice is often between maximising return from investment or minimising size of investment to achieve adequate performance.

The value relationship can be assessed as:

$$\text{Value} = \frac{\text{Measure of Satisfaction (benefits)}}{\text{Measure of Resources Consumed}}$$

which is often simplified in construction to:

$$\text{Value} = \frac{\text{Measure of how well a design delivers Function (requirements)}}{\text{Measure of project capital or whole life Cost (monetary)}}$$

It is important to understand the relationship between value and resources. Non monetary resources can be very important. The customer is prepared to pay a

---

[1] BS EN 1325-1: 1997 Value Management – Vocabulary, Standard, available from BSI London.

The Design Manager's Handbook, First Edition. John Eynon.
© 2013 The Chartered Institute of Building. Published 2013 by Blackwell Publishing Ltd.

market price for a commodity or service, but other resource attributes will also have an influence on the decision to purchase – core competences, delivery speed, delivery time, flexibility, quality of aftercare, reliability of the product, environmental impact, reputation of the organisation, etc., etc.

If a business intends to improve value, it needs to focus on delivering more of what the customer wants (performance) and do so while achieving better value for money in the supply chain. At the initial stage of a value study the focus is on agreeing value criteria, which govern decision making and establishing design requirements.

So *how can we maximise value? . . .*

There are national and international standards of practice that give guidance on approaches, methods and tools that can be used to recognise and manage value. Values that underpin the culture of an organisation combine with behaviours and skills to make a difference for the organisation, its people and its customers.

Value Management is defined in the standard **BS EN 12973**[2] as a style of management, particularly dedicated to motivating people, developing skills, promoting synergies and innovation, with the aim of maximising the overall performance of an organisation. Applied at the corporate level, value management relies upon a value-based organisational culture taking into account Value for both stakeholders and customers. At the operational level (project-oriented activities), it implies in addition the use of appropriate methods and tools.

Value Management can be thought of a framework that enables organisations to innovate and to plan and deliver value for money.

There are many frameworks in design and construction management. At a detailed level people may cite the time/cost/quality balance, but in reality, for any project, other factors too govern decision making.

The standard **BS EN 12973** illustrates the value-management framework in the context of corporate goals and introduces methods for use at working level. The Directors and Board should communicate clear policy and consistent guidance for the trade-offs that must be made to optimise return on investment. The high-level framework is about fit with corporate objectives/customer objectives and creating a value culture within the organisation, so that every decision at every level aims to add value. The standard framework (EN12973) also includes making sure all employees understand value, recognise training implications and give feedback to improve performance.

Organisations that set standards for practice achieve results. Value-management approaches covered by international standards deliver proven results. Benefits far exceed the cost of value-management effort: returns of order 100:1(!) and higher are regularly reported in official government statistics in the USA.[3] Best performance is delivered by maintaining a value culture and by applying skill and understanding of methods to enhance value in the existing environment and to create new opportunities for continuous improvement.

The basic approaches to value management can be readily learned by shadowing a practitioner or joining practical training, such as that advocated by The Institute of Value Management.[4] For those who become proficient and can demonstrate

---

[2] BS EN 12973: 2000 Value Management, Standard, available from BSI London.

[3] http://www.fhwa.dot.gov/ve/2010/ – Annual report for 2010 indicates return on investment 146:1!

[4] www.ivm.org.uk

effective results there is a route to professional certification, which is signified by the European qualification 'Professional in Value Management' (PVM).[5]

The Institute of Value Management (IVM) is a UK-based body that supports professionals in the field of value management. Other UK bodies (e.g. Association for Project Management and the Cabinet Office Best Management Practice programme) also identify value management as an integral part of professional development. The IVM, through its independent Certification Board, is responsible for implementing the European Certification and Training System within the UK. Value Institutions around the world may link with IVM.

What methods should Design Managers understand? . . .

Five core methods are described in this paper.

One key message is that Value Engineering is just one of five core methods.

The five core value-management methods are described in the standard (EN12973):

- Function analysis (about how things work in terms of what people/things do or need to do – analysis of situations and problems);

- Function cost (the money goes on <u>doing</u> what?!!!);

- Function Performance Specification (setting output specifications/ being clear about what the basic essentials are and what the performance objectives are – making sure everyone knows what their work enables others to do);

- Value Engineering/planning/analysis (optimising solutions, stimulating innovation (there are lots of creative thinking tools beyond brainstorming) – optimisation may be in terms of profit or other objectives);

- Design to Cost (sometime called the price minus approach – about delivering a product that meets the performance criteria by recognising the impact that every decision has on (total) cost: can be used to convey the benefits of spending more time and money on construction to achieve better business benefits in the long run for the client).

And how do these methods work? . . .

Value-management methods work by focusing design innovation exactly where it is required. Ask how else can we fulfil the function, rather than how can we find a cheaper product. It's a question of thinking about what things do for a customer and searching for better and more cost-effective solutions that meet the need, rather than just focusing on finding cheaper versions of the design product already selected. The Design Manager should add value by identifying better and more affordable solutions – not cheapen the design by buying less.

Value management assures productivity by its focus on value. Decisions focused on value should be right first time. Value thinking also eliminates waste in design and in construction activity and links with Lean. Lean thinking incorporated Value Engineering and developed from studies of manufacturing practice in the 1980s.[6]

Value management assures sustainability by incorporating all stakeholder views, and by enabling the team to consider environmental value criteria and community value criteria, as well as financial value criteria over the asset life cycle.

---

[5] http://www.ivm.org.uk/training-certification.php Certification Route.
[6] The Machine That Changed The World: Womack, Jones, Roos, Free Press Paperbacks, 1990.

Examination of function makes it plain what the requirements are and how these can be achieved. Function analysis is about understanding the connections and relative importance of all the various requirements that must be met and ways in which a design performs. Functions for a building project, for example, may be linked to the building performance (e.g. accommodate staff and increase productivity), the technical performance (e.g. transmit load and reduce emissions), and the contract performance (e.g. eliminate waste and simplify construction). The BCIS standard summary of elemental cost[7] can be a useful prompt to check the elements for the cost plan, but it is focus on function that enables alternative solutions and options to be developed.

The function cost method involves allocating costs in a design against each function that must be (or is) fulfilled. Inevitably some functions are more expensive to fulfil than others; and initially it may be discovered that some functions are fulfilled poorly or not at all. The Design Manager should examine these areas of underperformance, and the balance between expenditure on highly important functions and expenditure of less important functions. This method enables the team to reallocate resources to better effect and frequently frees up cash to accommodate design development.

Function Performance Specification is an extension of function analysis to develop specific and measured statements of requirement. Organisations depend on their capacity to provide competitive products that are best suited to satisfy the need of the users, whether expressed or implicit. The statement of the need in a functional form, i.e. in terms of purpose, without reference to solutions (technical, administrative, procedural, organisational, etc.), ensures that there is every chance that competitive and innovative solutions will emerge during design. This form of specification is also called an output specification – it's about what the design does or should do for the customer, rather than what the design is or should be. The design manager can use this form of specification to set performance requirements for the various design work packages and to assess suitability of designs delivered by subconsultants and subcontractors.

Value engineering involves following a standardised process sequence known as the job plan or work plan. Typically this involves a workshop. In some industries it is normal to run workshops over five days and use that time for concentrated effort on product design and development. Such an approach is not additional to routine design – it is about concentrating the routine design to one time in one location, so that the team can perform very efficiently and produce innovative, cost-effective results. In UK construction it is more likely to convene a workshop of only one or two days so the Design Manager is left with many decisions to make between value-engineering interventions. Value-engineering studies should be led by competent facilitators.

Design-to-cost involves setting up management systems that enable the Design Manager and the design team to understand the impact of changes before these are committed. Design development is then managed within a cost envelope. This method requires the Design Manager (and other team members) to set target costs for delivery of each function (as well as build cost). The method enables trade-off between functions and introduces control on decisions that must be made between higher-cost, high-performance design solutions and lower-cost and barely adequate design solutions.

---

[7] www.bcis.co.uk/downloads/Standard_Form_of_Cost_Analysis_-_Forms_1_.pdf

How is value management deployed?

Value-management methods are applied throughout the project life cycle in construction. Clients, professional advisers, and contractors use value management to analyse business requirements, make sustainable decisions, strengthen partnership teams, agree performance targets and prepare project specifications, as well as to improve value for money, save time, stimulate creative designs and streamline delivery processes. Value management delivers effective, efficient and economical projects.

The Design Manager can anticipate two types of value-management activity. There will be discrete studies at certain milestones, and there will be work to create an all-pervasive culture of value-based decision making. The Design Manager should make sure that there is a consistent focus on value priorities established for the customer's project across all individual organisations involved in the supply chain. One way to do this is to establish clear value criteria and prioritise them at the outset of any project. Establishing client project value criteria in a workshop environment with the whole team informs and establishes a productive project team.

Value study activity is anticipated throughout the project procurement life cycle[8]. In addition the standard (EN12973) expects that every decision should be value-based and that each organisation (or supply chain) should understand what the various value criteria are and what their relative importance is towards maximising the overall performance of the organisation

A significant challenge for industry is to apply value-management methods more effectively. Design Managers have a key role in this.

The focus of value management in UK construction is often only on value engineering late in the design period, and then sometimes only part of that approach is adopted – typically the brainstorming ideas session. This narrow view leads to short-term solutions that are project-based or cost cutting-based.

The 'standard' objective of value management is 'to maximise the overall performance of the organisation'. This is much wider than just taking that common, project-focused view. Cost cutting in one element of the overall value chain can generate consequences elsewhere. Typically costs are reduced by doing less or building to a lower spec, but this often generates extra costs under a separate budget heading (e.g. maintenance) or reduces income (e.g. low-carbon HQ developments attract better rent in the global market).

Some tips for getting started . . .

Effective value management involves coordination of design management activity and senior management activity.

Rather than doing less, can we find a solution that is lower-cost to build and delivers better performance in terms of profits for the supply chain and ongoing benefits for the customer?

Improving value could mean reducing cost without corresponding performance loss from the product, or doing something innovative at a different cost to achieve higher performance levels.

---

[8] Achieving Excellence Guidance : Project Procurement Lifecycle The Integrated Process OGC, 2007 http://webarchive. nationalarchives.gov.uk/20110822131357/http://www.ogc.gov.uk/documents/CP0063AEGuide3.pdf

Value management is *not* a workshop; it is the combination of day-to-day decisions and planned studies. The Design Manager should promote consistent, day-to-day decision making based on value.

Here are a few perceptions of value to consider:

- value in design (Design Manager: meeting the spec well and delivering a product that (we)/industry/client can build/lease/occupy and make a decent profit);

- value of design (Design Manager: business focus rather than project focus: our values, brand, ethos and priorities for long-term business success);

- Value management (approaches, standard processes, toolkit, etc.).

The Design Manager will be expected to demonstrate skills in applying value management. Accredited training courses are signposted by The Institute of Value Management.[9]

Some examples of what works . . .

Project Example: Recognise that value is about more than cost for civils contractor:

The client asked the contractor to take on ideas from the whole supply chain and recognise where small details could have a big impact. The key was to understand what the design was aiming to achieve (function analysis) and allocate cost to function. This highlighted the parts of the contract and design that could be improved. Simple things made a big difference – e.g. architectural features based on modular design combining standard elements or bespoke; and recognising the value of 'time'. The contractor's major contract team wanted to haggle for compensation around a small scope reduction valued at £300k. The remit was to 'reduce cost' and 'improve affordability within the cost target'. The team had not looked at time, as they thought that had no bearing on their budget, but by examining 'value and what drives 'value' across the supply chain it became readily apparent that speeding the work up was much more valuable than reducing scope to the customer. The purchaser then agreed to increase scope to speed up the job, increase profitability and reduce the overall cost to the purchaser. These activities were introduced by the client's 'Value Champion', with professional support to stimulate performance.

Project Example: Value management doubles Profit Margin for Regional Builder:

Every worker on a on a construction site was challenged to save a 'tenner'. All benefits initially accrued to the workforce, e.g. clocking off early, social fund benefits, etc., then benefits were applied to the building contractor company. The crucial element for success was to focus all creative thinking on what really adds value for the company. This ensured that one person's 'saving' did not just become another's 'cost' or delay. The value-management framework established the means to understand what drives value and communicated this through management workshops and toolbox discussion, so that sustainable ideas could be signed off by any authorised member of the workforce and implemented. In aggregate all ideas enabled 7 per cent of the project budget to be reallocated to fund an additional project. These activities were led by the building contractor project director, working closely with the design team and professional support to initiate activity.

---

[9] www.ivm.org.uk Institute of Value Management

Project Example: Value management improves sales success for Regional Builder:

Function analysis of client requirements focused creative thinking for a £20m tender. The client recognised that the builder really understood what drove value for the client. The builder was able to win on grounds of innovation in design and practical demonstration that their management methods controlled value for money and design management decisions with the client's interests at heart. By focusing creative thinking on what the client really needed, the winning alternative bid clearly demonstrated what the design did for the customer, and at the same time created the internal value culture and design to cost framework that assured profitability for the builder. These value activities were led by an experienced contractor's Design Manager with professional support

Project Example: Value management frees up working capital to fund marketing and growth:

A local authority had embarked on a long-term building refurbishment programme and used value thinking as a means to transform speed of delivery by planning for 'right first time'. Initial work with top-level management, then the whole supply chain together, established consistent understanding of what drives value for the organisation and how the supply chain could align to deliver that value. The increases in productivity enabled the team to recover a full year's backlog over a period of three years, without calling on extra resources. These activities were led by an independently chaired programme-management team formed of the client, two contractors, and the designers.

Project Example: Value management aligns partnerships for National House Building Programme:

Workshop activity and spreading a culture of consistent, value-based decision making aligned the team and motivated productive teamwork. The initial workshop enabled the team to understand what drove value for the partnership as a whole, and how that then impacted or was reinforced by each individual company's efforts to enhance commercial performance. The main contractor employed professional support to design this programme at the outset, then passed responsibility to the supply chain to deliver continuous improvement. These later activities were led by an independently chaired steering group of commercial managers from the principal subcontractors.

Project Example: Value management assists business sale:

A small, bespoke joinery and shop-fitting business (50 employees) was struggling with pressures of work for several competing and important customers. The owner wanted to sell the business too. Work was reactive and the company was in danger of losing money though inefficient spreading of workforce between contracts. Invoicing was often late and minor errors were being used by customers as excuses not to pay. Introduction of a value-management framework to cascade a value culture down from the top enabled all employees to understand what creates value for customers and the business. The team then reworked contract management and control procedures. Three immediate benefits were secured:

- a substantial cash-flow benefit, as outstanding invoices were paid and future payments were then received on time;

- a sound basis for prioritisation of resources was agreed, which enabled conflicts between contracts to be managed on profitable business terms;

- a due-diligence review of the business identified a company workforce that was well aligned to value for the business and as a result a fair premium price was negotiated for the business sale and all employees were retained!

These activities were led by a Design Manager with professional support, working in close liaison with the Managing Director.

Signposts for further reading . . .

1. www.ivm.org.uk: Institute of Value Management. Website contains general information, useful contacts, and a summary of the certification scheme for value-management practitioners. The key qualification to look for when buying value-management services is 'PVM', Professional in Value Management.

2. http://webarchive.nationalarchives.gov.uk/20110822131357/http://www.ogc.gov.uk/ppm_documents_construction.asp: Achieving Excellence Publications, OGC give general guidance on good practice.

3. *Value Management of Construction Projects*, John Kelly, Steve Male and Drummond Graham, Blackwell Publishing, 2004.

4. *Value Management Practice*, Michel Thiry, Project Management Institute, 1997.

5. *Value and Risk Management: A Guide to Best Practice*, Michael Dallas, Blackwell Publishing, 2007.

6. *Management of Value*, Michael Dallas and Stephanie Clackworthy, TSO, 2010.

CONTACT INFORMATION

Michael Graham

e-mail:	michael.graham@ukvaluemanagement.co.uk
Phone:	0797 442 1151
Web:	www.ukvaluemanagement.co.uk

Michael and his business colleagues offer one of the UK's largest qualified resources in value management. His business works throughout the UK and internationally.

APPENDIX E   Style, Behaviours & your 'Leadership Moments'

Saima Butt

Introduction

Two cars were being driven fast and in opposite directions along a winding country lane. It was late summertime, and the hedgerows on either side of the lane were lush and high. It was impossible to see around any of the corners.

Both drivers, because of the heat of the day, had their windows wound down, and their minds focused on the road ahead. And, as it happened, the driver of one car was a man and the driver of the other was a woman.

They approached the final bend at speed, and they only just managed to see each other in time. They stood on their brakes, and just managed to slide past each other without scraping the paintwork.

As they did so, the woman turned to the man, and through the open window she shouted, 'PIG!'

Quick as a flash he replied, 'COW!'

He accelerated around the corner . . . and crashed into a pig!!!

Lionel Blue, BBC Radio 4 'Today' programme

Individual Styles & Behaviours

This section briefly introduces the impact of our cognitive style on our everyday life. Cognitive style refers to our thinking and how we go about doing what we do in our own preferred way to get things done. This unique thinking style becomes, over time, our comfort zone for how we choose to do things – our personality.

Personal style therefore has a significant effect on how we communicate to our self, relate with each other and go about our daily activities of problem solving, making decisions and being creative to deliver results.

At times when we are not working within our own comfort zone, some tasks or decision making may need slightly more effort and thought. This subtle difference to style becomes our behaviour.

How well we cope with this difference in behaviour depends on how clear and tangible our outcomes and goals are and how motivated we are to achieve them.

To understand more about our style we can reflect on the following two lists and how you might prefer to operate . . .

Some might . . .	while others . . .	. . . and how about you?
Think outside the box	Think inside the box	
Do–Think–Do	Think–Do–Think	
Non-linear	Linear and methodical	
Big chunk	Small chunk	**?**
Do it differently	Do it better	
Challenge, reframe	Work within framework	
At ease with ambiguity	Require order and structure	
Big picture	Enjoy the detail	

The Design Manager's Handbook, First Edition. John Eynon.
© 2013 The Chartered Institute of Building. Published 2013 by Blackwell Publishing Ltd.

What you might notice is that you can do both lists and implicitly know what you prefer. So labelling yourself up as one style may potentially limit your behaviour or how you get the job done. Being motivated and having clear sight of the outcome or goal means that you can adapt (incrementally) and innovate (transformationally), to perform at your best and deliver.

Leadership Moments

These are opportunities when we just know what we want, feel it and connect with what's right for us and those around us. It is these moments that make us successful. How we choose these moments can therefore have an impact on our professional career – not to mention our personal life!

As you look through the bubbles in the mind map, pay attention to how you operate now and consider how you could behave in the future to get better results. Begin to notice what comes easily to you – and perhaps what doesn't. What are you always already doing and what might you benefit from learning to do more of? What's missing in your style?

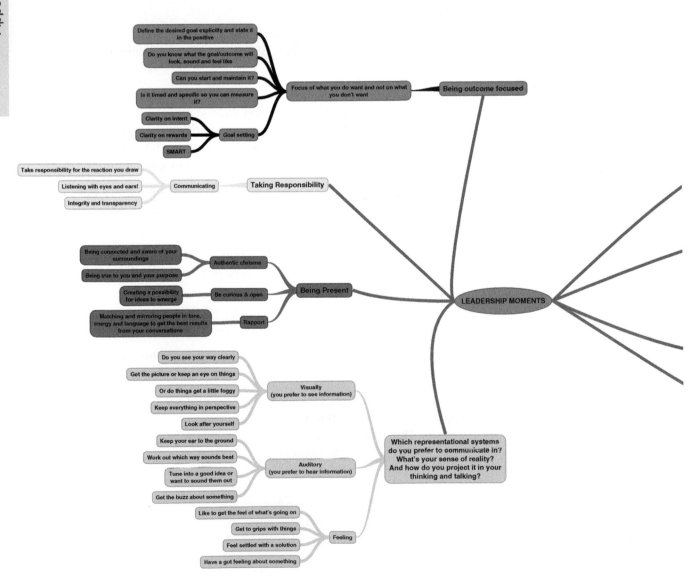

These questions are designed to raise your awareness of how you understand yourself, and so may create more flexibility in your style and behaviours. It is an art and not a science, so just be curious as to what strikes you and how you may benefit from the new learning.

The most successful teams are goal-centric, with a heightened awareness of each other's styles, and they know best how to deploy them – right people, right place, right time!

Summary

Taking time to reflect on your own style of leadership and how it may affect others and the end objective is key to any professional development. Simply relying on how things used to be done and your experience of having done it that way may not be viable in an ever-changing world.

Adopting openness towards change can become one of the first steps to even more effective leadership of you and your teams. So, for you, what's the difference that makes the difference? What do you notice about how you might do things differently now? Take a moment and reflect on your style and consider:

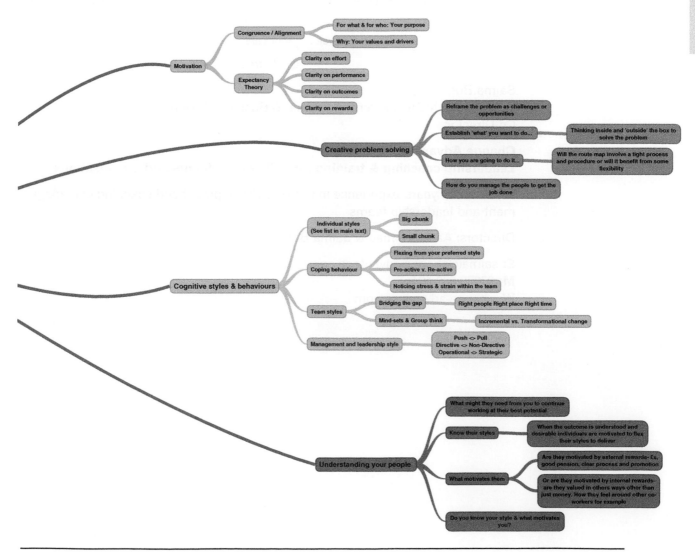

- How does your style impact the job you are doing?

- What assumptions do you make about your style?

- How can you adapt your style to suit the job at hand?

- What are the assumptions others might make about you?

- How can you integrate others better to get the best results?

So be open to behavioural flexibility for the best results, or you may just hit a rock…

> *An armada of the US Navy was engaged in naval exercises off the coast of Canada when the following radio exchange was recorded:*
>
> #1 *'Please divert your course 15 degrees to the north to avoid danger of collision.'*
>
> #2 *'Recommend that you divert your course 15 degrees to the south to avoid danger of collision.'*
>
> #1 *'We repeat. Divert north now to avoid collision.'*
>
> #2 *'Strongly recommend you divert south soonest to avoid mishap.'*
>
> #1 *'This is the Captain of a US Navy warship. I say again, divert your course with immediate effect.'*
>
> #2 *'Copy. We say divert south now.'*
>
> #1 *'This is the USS* Enterprise. *We are an aircraft carrier of the US Navy. Divert your course NOW!'*
>
> #2 *'We are a Canadian lighthouse . . . Your call!'*

– Quote from *The Magic of Metaphor*, Nick Owen

Saima Butt
BSc (Hons), MBA, Master Practitioner & Business Coach
Co-Director

Change Advantage
Leadership coaching & training for individuals, teams and organisations

With over 10 years' experience in personal development and coaching of management and leadership teams

Directors: Aamir Ahmed & Saima Butt

E: saima@change-advantage.com
M: 07973 632 171
W: change-advantage.com

APPENDIX F Educating the Design Manager of the future

Paula Bleanch

If we imagine the Design Manager of the future, perhaps in ten years' time, who will that person be, what will they do, and how will they have progressed into the role? This section will consider the issues of educating the Design Manager of the future.

The Professional Bodies, Industry and Academia are the three key stakeholders involved in this process. From discussions that have taken place between these groups in the context of the Design Management Forum and Working Group, it would seem that each has similar ideas about what will be required from graduates and the sort of skills they should possess, but it appears that we do not have a good understanding of what everyone should bring to the party. During these discussions it became clear that there were misunderstandings among the groups about what everyone should be doing, what they are doing and what we can all do to achieve the outcomes we desire.

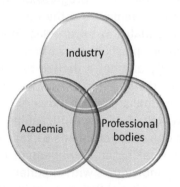

The relationship between the key stakeholders in Design Management education is vital. We need to agree where the 'overlaps' are and pass the baton between these bodies more effectively.

It is the role of the professional bodies to represent us and make jobs in the industry attractive to bright, talented young people. They need to ensure that potential Design Managers decide to join construction by selling this as an interesting and engaging career choice. It's generally accepted that construction on the whole has a poor reputation, and it's vital that we try to overcome this. It would be great if we could finally get over the 'builder's bum' perception of construction, and attract a more diverse range of young talent into the industry at a professional level. The media is to a certain extent to blame for the situation, as there is biased coverage of 'cowboy' builders and architects, but not much in between. The full range of professional routes, people who work daily on construction projects, are rarely seen by the public. New construction courses at GCSE level are designed to encourage school pupils to become aware of the 'hidden' professions in construction; hopefully these curricula will help pupils to understand the plurality of construction professions and consider these as possible careers.

The professional bodies accredit degree courses provided by universities, and then inspect them to ensure that they come up to their required published standards.

The Design Manager's Handbook, First Edition. John Eynon.

213

Appendix F

This means that graduates can progress sooner to full professional membership after graduating from an accredited course. This accreditation, to a large extent dictates the content of courses, because universities have to cover the requirements in their teaching in order to maintain their status and attract students onto the course. This suggests a dynamic between the professional bodies and academia, which should be informed by the requirements of industry as the end user. At the moment the professional bodies are re-orientating themselves to their membership and potential members, making changes to the way they recognise what people actually do in practice, and Design Management is one key area. As there are only three undergraduate degree courses in the UK, the majority of Design Managers are educated under other umbrellas and develop into the role in practice. These Design Managers want to have some recognition of their specific skills and it is right that the professional bodies should provide that. For some professions, like Architectural Technology, design management as a process may be part of the role, although the individual may not call themselves a Design Manager.

Having attracted young people to join the industry, what do we then do with them? In the past the career progression in construction was through working on site, becoming a foreman and then a manager. Today most construction companies employ graduates in these roles. Even if someone has worked their way up to a management position, they often find that they are expected to return part-time to university and get a degree to achieve parity with their graduate colleagues. This poses the question of how we value degree education and what value it adds to an individual and to the industry. Is the industry any better now than in the past, when we relied on apprenticeships and individual aptitude to rise to the top? Perhaps one of the problems we face is that we have graduates with very little site experience taking up management roles in construction. Industry always wants to employ individuals with experience; however, at the same time industry is less than forthcoming at the moment in offering placement work to students who want to work on site. Perhaps this is one of the areas where the baton is currently being dropped?

In terms of the role – and after all, Design Management is one role definitely invented by industry to fulfil an industry need – how likely is it to remain static and therefore easy to describe and accredit (if this is our goal)? If we look at the change in the tasks that quantity surveyors (QSs) are expected to undertake as part of their role now in comparison to 15 years ago, we see very few QSs today who actually write Bills of Quantities. The shift has been driven by the changes to the procurement routes we use and it reflects on the other professions. Leadership of the design team has moved from the architect to the contractor, to reflect contractual relationships. Specialist subcontractors who design and build are more common owing to changes in technology. This implies that whichever profession we regard ourselves to be, flexibility is more and more important and that perhaps puts more emphasis on wider skills and on understanding the process. It would suggest that soft skills, transferable skills and, most importantly, team-working skills are needed to be able to undertake the role now and in the future. By simulating real projects in university, we can perhaps 'future-proof' graduates to some extent, by ensuring that they leave university with a picture of what is happening and what may happen.

Ultimately a tension will always exist in construction education over whether we are training people for their role in industry, or educating them for their future in construction. Perhaps it should be clear that we are doing a bit of both?

	Professional Bodies	**Academia**	**Industry**
Education	Accreditation of academic qualifications. i.e. deciding what universities need to teach to fulfil industry demands	Provision of suitable courses that fulfil the accreditation requirements of the professional bodies. Keeping at the forefront of research and introducing this into the degrees to keep informing industry.	Wants to employ graduates with certain skills and abilities and rely on the professional bodies and academia to produce them.
Training	CPD courses to ensure members are maintaining currency	Training in some generic tools required to do the job, in terms of software and ways of working.	Ensuring graduates have the right training to fulfil their expected job role within their organisation.

Summary table of the expectations of each of the key stakeholders in providing education and training

A question remains over whether the different bodies are communicating adequately in the creation of the Design Manager of the future. What should we be doing? How do we 'hand over the baton' from stage to stage through the development process of the Design Manager? What value does each of the stakeholders add at each stage in the process?

Perhaps we can gain an insight into what is currently happening through research recently conducted by Brendan Keilthy for the FAS scholarship in 2010. He asked graduates from the three Design Management courses (at Greenwich, Loughborough and Northumbria Universities) what they thought of their education, and then asked industry to comment on what they thought of graduate educations. The results were very interesting.

Industry told him that the following key skills required improvement:

• Understanding the design process

• Leadership

• Technical knowledge

• Ability to understand drawings.

Industry also identified that graduates could benefit from more site experience and by covering the following subjects in more detail:

• Building Services

• Construction technology

• Sustainability

• Design management

• Building materials . . .

. . . whereas these subjects could be covered less or not at all:

- 2D/3D CAD

- Economics and finance

- Surveying and Law.

A number of missing themes were identified by the respondents, including:

- Structural design

- Building Information Modelling (BIM) (seems to contradict the finding above?)

- The necessity for site experience

- Planning and programming

- The range of design processes and associated approvals

- Design team appointments / responsibilities

- RFIs

- Extranet management

- Building regulations

- Contractual awareness (different from law (above) in what way?).

The findings above seem to suggest that students are lacking knowledge about how design is managed in practice and also how the other members of the team work. On this basis, should we adjust our view of what is important for student Design Managers to learn in university? Industry will often question the practical training of graduates, and universities may say that some things are better learned on the job. Maybe that is where the placement experience becomes key, and by collaborating better over that experience, could we perhaps close the gap? However, this also works in reverse. Take for instance BIM, where universities recognised the need for teaching and research in this area before industry did, so graduates have left with these skills, built up knowledge in practice and now use their skills on projects, coming back into universities to teach the next generation of students.

Thinking about the desired skills and knowledge of graduates should also make us question who is teaching them. Is this also forcing us towards more collaboration between industry and academia in the classroom? Should students be taught by practitioners working part-time in education, or should lecturers keep working part-time in industry to maintain their currency? Either route would ensure that educationalists kept up to date with progress in industry and could hopefully mean that their research is industry-orientated and practical, ultimately fulfilling industry's needs.

If we consider what industry actually does, the survey provided us with an interesting insight. Only 20 per cent of the respondents said that their company employed Design Management graduates *and* had specific training for them. Less than 15 per cent of respondents thought that specific training was 'not important' or only 'somewhat important'. Clearly there is a gap here that, between us, we need to fill. Industry needs help from academia and the professional bodies to do this training; it is unlikely that they will be able to provide it via their individual graduate training schemes. On this point I think back to my own experience as a graduate, being given the job of Design Co-ordinator on a large project and being asked to write down my job description, since my company liked what I was doing, thought I was doing a good job, but just 'wasn't sure what it is you actually do'! Perhaps this is where industry has to pull its socks up. How many of us know of people who have been recruited into Design Management who aren't suited to the job, don't understand the design process and thus give Design Managers a bad name? Design is one of the UK's best-selling exports; it deserves to be managed well, but designers have been much criticised over the past ten years or so for not understanding enough about building. I would turn that around and say that it's about time builders understood more about the design process, especially now they are in charge of it on D&B projects.

One of the outcomes of Keilthy's research was to identify a gap between what graduates do and what is covered in their education. This included the following tasks:

- Update drawing registers

- Manage change

- Submit red-line drawings

- Manage the approvals process.

It also outlined some key areas to target:

- Improving technical skills (ability to understand and interpret technical drawings and specs)

- Cover design approval processes, designer's appointments, extranets, typical contractual arrangements

- Introduction of workshops that stimulate typical DM day-to-day activities.

Which of these should we include in education and which in training? Perhaps we need to have a really honest dialogue about this. We also need to consider that with the increasing use of BIM on projects, many of the tasks outlined above will still need to be managed; however, they will not be physically being done by Design Managers using 2D drawings, as BIM will potentially be doing that for us.

In academia we have a lot of pressures and much more to come with the changes to fees. We don't actually have a lot of time to teach (24 weeks a year), so perhaps it's more important for us to give students the ability to teach themselves in the future. The key is to agree with industry and the professional bodies on where the boundaries are and how we are communicating to students what they are required to do to perform the role.

An important question is whether undergraduate level is the right place to teach Design Management. After all, it's pretty difficult for someone who has not experienced designing and building to understand what a Design Manager does. In terms of theory, all the literature on design management is relatively new (having been produced only in the past 15 years or so). It is a hard topic for undergraduates with little experience to understand. (Here I rewind several years to conversations I had with potential employers about how I'd have to serve quite a bit of time on site before they'd let me near a design team.)

Perhaps one solution is that we do enough in undergraduate courses to cover the basics and then leave the rest to master's level/CPD/industry experience, or a combination of all three? One thing is certain: that there is no one correct route to design management or becoming a design manager, and in my opinion that diversity of knowledge and talent can only be good for the profession.

So what do we want from the design manager of the future? Isn't it about time that we got together and had a chat about it, and decide how we are going to produce that person and what they should be?

Paula Bleanch
BSc (Hons), PGCERT, MSt. IDBE, ICIOB
Senior Lecturer

+44 (0)191 227 3213
paula.bleanch@northumbria.ac.uk
http://www.northumbria.ac.uk/sd/academic/bne/study/aec/acestaff/paulableanch
Twitter: @paulableanch
LinkedIn: http://uk.linkedin.com/pub/paula-bleanch/23/11a/a36

Biography

I took up my current role as Senior Lecturer in the School of the Built and Natural Environment in October 2008. I teach students on the Construction Related Programme, consisting of BSc (Hons) Construction Management, Project Management, Design Management and Architectural Technology. The ethos of the degree course is that each route has its own specific professional bias, but the students are taught together to encourage the development of teamwork skills and an awareness of the other roles in the construction and design processes.

I myself graduated from the BSc (Hons) Building Design Management degree at Northumbria in 2003, following a few years working for a large contractor and studying architecture. The Design Management course altered my career expectations, especially as I had the opportunity to spend my placement year in Paris working for a project management consultancy, which was an amazing experience. When I began the course at Northumbria, I finally felt as though I had found my 'home', as someone who was interested in both design and the management and building of projects, I had always struggled to see how I personally fitted into the big picture. After starting the course I finally stopped feeling like a square peg in a round hole!

After graduation, I worked on site as a Design Manager for several large contractors on big public/private partnerships, building hospitals, schools and offices. I also worked for a well-known local architectural practice before joining the teaching staff at Northumbria. Projects that I am particularly proud of are Newcastle City Library and Darlington Education Village.

I'm passionate about design, construction and improving the way we work in the industry. I manage a module teaching BIM to Construction, Project and Design Management for students, which we think is unique in the UK. I also teach technology, the final-year design management module, and I co-ordinate several staff working on the final-year interdisciplinary project where all students work together in mixed groups.

Research interests

I am interested in the people issues and intangibles that make construction projects so interesting. I have recently completed a part-time Master's degree in Interdisciplinary Design for the Built Environment at the University of Cambridge, looking at how professionals can work together more effectively within teams.

Key publications

Bleanch, P. (2010) 'A case study investigation of contractual and relational governance in PFI'. Unpublished Master's thesis, University of Cambridge.

Braithwaite, J. and Bleanch, P. (2011) 'The Architectural Technologist's role in integrated design'. Awaiting publication.

Other information

I'm on the judging Panel for the Constructing Excellence North East Awards and have sat on the NBS Advisory Panel as their academic member.
In my 'spare' time I'm attempting a PassivHaus refurbishment of our home with my partner David.

APPENDIX G   Facilitating workshops

John Eynon

'The Rules'

I came across these several years ago, and have used them as a basis for facilitating workshops when appropriate, just to encourage the right kind of thinking and get the group expectations and behaviours set at the right level.

DO:

- **Stay loose and fluid in your thinking**
- **Protect new ideas from criticism**
- **Acknowledge good ideas, listen, show approval**
- **Eliminate status or rank**
- **Be optimistic**
- **Support confusion and uncertainty**
- **Value learning from mistakes**
- **Focus on the good aspects of an idea**
- **Share the risks**
- **Suspend disbelief**
- **Build on ideas**
- **Avoid evaluating too early; delay resolution**
- **Open questions**
- **What if?**
- **Reject detail at this stage**
- **No taboos or sacred cows.**

DON'T:

- Interrupt, criticise
- Be competitive
- Mock people
- Be dominant
- Disagree, argue, challenge
- Be pessimistic
- Point out flaws
- Be inattentive or fail to listen
- React negatively

The Design Manager's Handbook, First Edition. John Eynon.
© 2013 The Chartered Institute of Building. Published 2013 by Blackwell Publishing Ltd.

- Insist on the facts

- Give no feedback, or act in a non-committal fashion

- Pull rank

- Become angry

- Be distant

- Ask closed questions

- Demand detail, practicalities

- Erect fences, walls, boundaries

- Have no-go areas.

– With thanks to Guy Huggins, MBA, while at Carillion PLC
(Guy Huggins – En-able Management Consulting Ltd – www.enablemcl.com)

Six Thinking Hats

(Edward de Bono (2000) *Six Thinking Hats*, Penguin)

Something else you might find interesting is the 'Six Thinking Hats' method. Edward de Bono is a well-known writer and thinker. Here he puts forward a method for structured thinking around issues, enabling all views to be considered and emotions to be released and channelled, if that's needed. It's a simple but effective way of structuring workshops that is different, and will provoke some changed thinking.

For the full detail you need to read the book itself – it's a relatively short, easy read. Good luck!

'The Six Thinking Hats method may well be the most important change in human thinking for the past twenty-three hundred years' – Edward de Bono, Preface, 'Six Thinking Hats'

Problem Solving and Filtered Thinking – Six Hats

Thinking about thinking . . .

We face and deal with problems every day.

We work in a solutions industry, and problem solving is what we do every day.

We use emotions, logic, information, hope and creativity, usually intuitively and may not realise it!

Could we be more efficient, use less time, create more value, in solving the problems we face?

We hope so!

The use of the six hats offers a way to structured problem solving, and enables us to step outside our preferred styles in a structured way.

Something to think about!

Bibliography

Design management reading list and further resources

With my thanks to Stephen Emmitt, Alec Newing, Saima Butt, Michael Graham and others . . .

The purpose of this reading list is to get you thinking and perhaps lead to some further research and learning.

There's a lot of other work out there on Design Management, not just in the Built Environment sector. Some of these publications I have already referred to.

In addition there's a lot of work going on in the academic field, particularly around lean construction principles and the impact and development of BIM technologies.

And one more thing: I've thrown in a few 'left field' books that I and/or others have found useful and that might give you some fresh insights and ideas.

Design management

Best, K. (2006) *Design Management: Managing design strategy, process and implementation*, AVA, Lausanne.

Best, K. (2010) *The Fundamentals of Design Management*, AVA, Lausanne.

Boyle, G. (2003) *Design Project Management*, Ashgate, Aldershot.

CDM (2007) *Approved Code of Practice*, HSE.

Cooke, B. and Williams, P. (2004) *Construction Planning, Programming and Control* (2nd edition), Blackwell Publishing, Oxford.

Cooper, R. and Press, M. (1995) *The Design Agenda: A guide to successful design management*, Wiley, Chichester.

Cooper, R., Aouad, G., Lee, A., Wu, S., Fleming, A. and Kagioglou, M. (2005) *Process Management in Design and Construction*, Blackwell Publishing, Oxford.

Emmitt, S. (2007) *Design Management for Architects*, Blackwell Publishing, Oxford.

Emmitt, S., Prins, M. and Otter, A. (Eds.) (2009) *Architectural Management: International research and practice*, Wiley-Blackwell, Chichester.

Gilbertson, A. (2007) *CDM2007 – Workplace 'in use' guidance for designers*, CIRIA.

Gray, C. and Hughes, W. (2001) *Building Design Management*, Butterworth-Heinemann, Oxford.

Bibliography

Ove Arup + Partners, updated Gilbertson, A. (2007) *CDM2007 – Construction work sector guidance for designers* (3rd edition), CIRIA.

Sinclair, D. (2011) *Leading the Team: An architect's guide to design management*, RIBA Publishing.

RICS (2011) Draft Guidance Note – Managing the design delivery, https://consultations.rics.org/consult.ti/managingthedesigndelivery/consultationHome

Tunstall, G. (2006) *Managing the Building Design Process* (2nd edition), Butterworth-Heinemann, Oxford.

Related Publications

Chappell, D. and Willis, A. (2005) *The Architect in Practice* (9th edition), Blackwell Publishing, Oxford.

Claire, J. (2010) *The Future for Architects?*, Building Futures, RIBA.

Dalziel, B. and Ostime, N. (2008) *Architect's Job Book* (8th edition), RIBA Enterprises, London.

Terry, A and Smith, S. (2011) *Build Lean: Transforming construction using lean thinking*, CIRIA.

DM related Handbooks/Codes

CIC Scope of Services:
The CIC Scope of Services Handbook (2007), RIBA Publishing.
The CIC Scope of Services CIC/Services (2007), RIBA Publishing.
The CIC Scope of Services CIC Consultants' Contract Conditions (2007), RIBA Publishing.

Dalziel, B. and Ostime, N. (2008) *Architect's Job Book* (8th edition), RIBA Enterprises, London.

Halliday, S. (2007) *Green Guide to the Architect's Job Book* (2nd Edition), RIBA Enterprises, London.

RIBA Outline Plan of Work (2007) RIBA Publications, London.

RIBA Plan of Work: Multi-disciplinary Services (2008) RIBA Publications, London.

Soft Landings Framework: pdf available from www.bsria.co.uk, BSRIA.

A Design Framework for Building Services (2nd edition), (BG6/2009), www.bsria.co.uk, BSRIA.

BIM (Building Information Modeling)

AIA Documents:
(2007) *Integrated Project Delivery: A Guide – Version 1*
(2008) Document E202 – *Building Information Modeling Protocol Exhibit*
(2008) Document C195 – *Standard Form Single Purpose Entity Agreement for Integrated Project Delivery*
                    (All available from the American Institute of Architects)

BS1192: 2007 – *Collaborative production of architectural, engineering and construction information* – Code of Practice

BSI/BIS (2010) *Investor Report – Building Information Modelling*, BSI.

CIOB CRI (2011) 'UK Government projects to use BIM by 2016: It's official', CIOB magazine article, September.

Crotty, R. (2012) *The Impact of Building Information Modelling: Transforming construction*, Spon Press.

Eastman Teicholz Sacks Liston (2011) *BIM Handbook: A guide to Building Information Modeling for owners, managers, designers, engineers and contractors*, Wiley.

Hardin, B. (2009) *BIM and Construction Management – Proven tools, methods and workflows*, Sybex/Wiley.

Richards, M, (2010) *Building Information Management: A standard framework and guide to BS1192*.

Sacks, R., Associate Professor, Virtual Construction Lab, Israel Institute of Technology, *KanBIM Project*.

Salford SCRI Forum (2011) *What do Contractors get from BIM and How?*

Smith, D.K. and Tardif, M. (2009) *Building Information Modelling: A strategic implementation guide*, Wiley.

WSP and Kairos Future (2011) *Ten Truths about BIM*: www.wspgroup.com/en/wsp-group-bim/10-truth-bim/

UK Government Reports

BiM (2011) *Management for Value, Cost and Carbon improvement: A report for the Government Construction Client Group*, March, Dept of Business Innovation and Skills.

HM Government, Carbon Plan, April 2011.

HM Government, Construction Strategy, May 2011.

Innovation and Growth Team (2010) *Low-Carbon Construction*: Final Report.

Value Management

British Standards Institution: British and European Standard BS EN 12973, *Value Management*, BSI, London.

British Standards Institution: PD 6663, *Guidelines to BS EN12973: Value Management – Practical guidance to its use and intent*, BSI, London.

British Standards Institution: British Standard BS 8534: 2011 Construction Procurement Policies, Strategies and Procedures – Code of Practice, BSI, London.

Dallas, M. (2007) *Value and Risk Management: A guide to best practice*, Blackwell Publishing.

Dallas, M. and Clackworthy, S. (2010) *Management of Value*, TSO.

Hogan, C. (2000) *Facilitating Empowerment*, Kogan Page.

Institute of Value Management: www.ivm.org.uk (Website contains general information, useful contacts, and a summary of the certification scheme for value-management practitioners. The key qualification to look for when buying value-management services is 'PVM', Professional in Value Management.)

Kelly, J., Male, S. and Graham, D. (2004) *Value Management of Construction Projects*, Blackwell Publishing.

OGC: http://webarchive.nationalarchives.gov.uk/20110822131357/http://www.ogc.gov.uk/ppm_documents_construction.asp Achieving Excellence Publications. (OGC give general guidance on good practice.)

Bibliography

Thiry, M. (1997) *Value Management Practice*, Project Management Institute.

(*Note that by June 2012 it is expected that a set of competence standards published by the various European Value societies will be available.*)

People

Austen, J. (1813) *Pride and Prejudice*, Penguin, London.

De Bono, E. (1998) *Simplicity*, Penguin, London.

Gallwey, T. (1986) *The Inner Game of Tennis* (2nd edition), Pan Books, London.
(2000) *The Inner Game of Work: Overcoming mental obstacles for maximum performance*, Orion, London.

Goleman, D. (2006) *Social Intelligence: The new science of human relationships*, Hutchinson, London.

Hargrove, R.A. (1995) *Masterful Coaching*, Jossey-Bass Pfeiffer, San Francisco.

Kirton, M. (2003) *Adaption – Innovation: In the context of diversity and change*, Routledge, Hove, UK.

Knight, S. (2002) *NLP at Work – Neuro Linguistic Programming: The difference that makes the difference in business* (2nd edition), Nicholas Brealey Publishing, London.

Owen, N. (2001) *The Magic of Metaphor: 77 stories for teachers, trainers and thinkers*, Crown House, Carmarthen, Wales.

Schutz, W. (1994) *Profound Simplicity* (4th edition), WSA, Mill Valley, CA.

Syed, M. (2010) *Bounce*, Fourth Estate, London.

Whitmore, J. (2010) *Coaching for Performance* (4th edition), Nicholas Brealey Publishing, London.

Other Stuff you might find interesting . . .

Arup, C.L. and the Arup Foresight team (2009) *Drivers of Change*, ed. J. Greitschus, Prestel Verlag: www.driversofchange.com

Collins, J. (2001) *Good to Great: Why some companies make the leap . . . and others don't*, Random House.

De Bono, E. (2000) *Six Thinking Hats*, Penguin.

Gladwell, M. (2000) *The Tipping Point: How little things make a big difference*, Abacus.

Godin, S. (2005) *Purple Cow*, Penguin.
(2008) *Tribes: We need you to lead us*, Piatkus.
(2011) *Poke the Box*, The Domino Project.
(2011) *We are all weird*, The Domino Project.

Mackenzie, G. (1998) Orbiting the Giant Hairball, Penguin Viking.

Moore, G.A, (1998) *Crossing the Chasm* (2nd edition), Capstone.

Peters, T. (2004) *Re-imagine! Business excellence in a disruptive age*, Dorling Kindersley.

Pirsig, R. (1970) *Zen and the Art of Motorcycle Maintenance*, Random House.

Sinek, S. (2009) *Start with Why*, Penguin Portfolio.

Taleb, N.N. (2008) *Black Swan: The impact of the highly improbable*, Penguin.

The Mind Gym (2005) *The Mind Gym: Wake up your mind*, Time Warner Books.

Turner, G. (1998) *The Personality Compass*, Element.

Tushman, M.L. and O'Reilly, C.A. (2002) *Winning Through Innovation: A practical guide to leading organizational change and renewal*, Harvard Business School Press.

Welch, J. and Welch, S. (2005) *Winning: The ultimate business how-to book*, Harper.

Papers

Ballard, G. (1999) 'Can pull techniques be used in Design Management?', *Conference on Concurrent Techniques in Construction*, Helsinki.

Ballard, G. (2002) 'Managing work flow on design projects: A case study', in *Engineering, Construction and Architectural Management*, 2002, 9.

Koskela, L., Huovila, P. and Leinonen, J. (2001) 'Design Management in building construction: From theory to practice', *Journal of Construction Research*, Vol. 3, No. 1.

Koskela, L., Ballard, G. and Tanhuanpaa, V.P. (1997) 'Towards Lean Design Management', IGLC-5 Proceedings.

Sir Ian Dixon and the CIOB Faculty of Architecture and Surveying Scholarship and other papers, sponsored by the Chartered Institute of Building and the Worshipful Company of Constructors:

2008/2009
Lisa Gould – Wates Construction:
'Can BIM Reduce Risk and Create Opportunities in a Declining Market?'
Robert Thompson – Interserve:
'How do Design Managers in major contractors generate, capture and assess 'build-ability' innovations and what barriers do they face'

2009/2010
Brendan Keilthy – VINCI Construction UK:
'Design Management Graduates: Fit For Purpose?'

2010/2011
Richard David – Willmott Dixon:
'Procurement and Contract Choice: A contributor to project failure'
Alistair O'Reilly – Laing O'Rourke
'BIM: A Waste Minimisation Tool?'

Articles

Chevin, D. (2011) 'Let's hear it for the Design Manager', *Construction Manager*, May.

'Design Management – an evaluation', *CIOB CRI magazine*, June 2011.

Journals

- Architectural Engineering & Design Management

- Architects Journal

- Building Design

Bibliography

- Building Research & Information
- Construction Management & Economics
- Construction Manager
- Construction Research and Innovation
- Design Studies
- Detail
- Engineering Construction and Architectural Management.

Index

The Design Manager's Handbook, First Edition. John Eynon.
© 2013 The Chartered Institute of Building. Published 2013 by Blackwell Publishing Ltd.

Index

Index

Printed and bound by CPI Group (UK) Ltd, Croydon, CR0 4YY

27/10/2024

14580289-0004